Studies in Computational Intelligence

Volume 1270

The series "Studies in Computational Intelligence" (SCI) publishes new developments and advances in the various areas of computational intelligence—quickly and with a high quality. The intent is to cover the theory, applications, and design methods of computational intelligence, as embedded in the fields of engineering, computer science, physics and life sciences, as well as the methodologies behind them. The series contains monographs, lecture notes and edited volumes in computational intelligence spanning the areas of neural networks, connectionist systems, genetic algorithms, evolutionary computation, artificial intelligence, cellular automata, self-organizing systems, soft computing, fuzzy systems, and hybrid intelligent systems. Of particular value to both the contributors and the readership are the short publication timeframe and the world-wide distribution, which enable both wide and rapid dissemination of research output.

Indexed by SCOPUS, DBLP, WTI AG (Switzerland), zbMATH, SCImago.

All books published in the series are submitted for consideration in Web of Science.

Rajan Gupta · Anamika Gupta ·
Sarabjeet Kaur Kochhar · Saibal K. Pal

Machine Unlearning: Concepts, Techniques and Applications

Springer

Rajan Gupta
Universidad Autonoma de Tamaulipas
Ciudad Victoria, Mexico

Anamika Gupta
Shaheed Sukhdev College of Business Studies (SSCBS)
Delhi University
New Delhi, Delhi, India

Sarabjeet Kaur Kochhar
Indraprastha College for Women (IPCW)
Delhi University
New Delhi, Delhi, India

Saibal K. Pal
New Delhi, Delhi, India

ISSN 1860-949X ISSN 1860-9503 (electronic)
Studies in Computational Intelligence
ISBN 978-3-032-18446-7 ISBN 978-3-032-18447-4 (eBook)
https://doi.org/10.1007/978-3-032-18447-4

This Springer imprint is published by the registered company Springer Nature Switzerland AG
The registered company address is: Gewerbestrasse 11, 6330 Cham, Switzerland

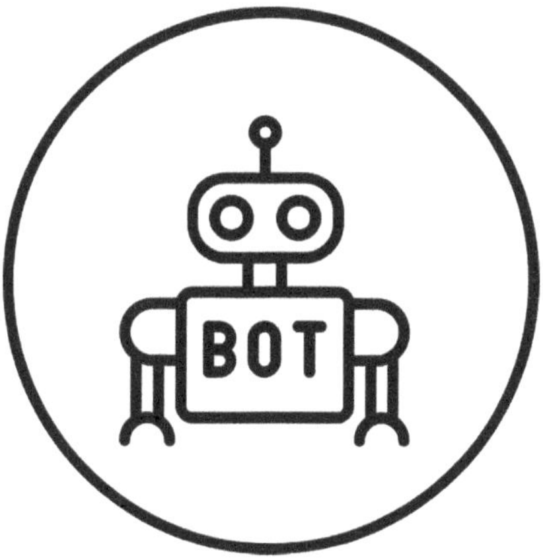
BOT

Dedicated to
*My **"Gurumaa"** for holding my hand in Life,*
*My **"Parents"** for making me stand in Life,*
*My **"Nephews—Reyaansh and Atharva"** for making me strong in Life,*
*My **"Brother and His Wife"** for helping me progress in Life, and*
*My **"Wife"** for supporting and loving me unconditionally in Life!*

***—Dr. Rajan Gupta**, Visiting Sr. Researcher, Artificial Intelligence and Innovation Lab, Tamaulipas Autonomous University (UAT)*

*My **"Parents"** for being my roots and wings,*
*My **"Husband—Ravi Gupta"** for being an unwavering support,*
*My **"Children—Arunima and Raman"** for being constant cheerleaders.*

***—Dr. Anamika Gupta**, Professor, Department of Computer Science, Shaheed Sukhdev College of Business Studies, University of Delhi, India*

*My **"Parents"**, who built my foundation and taught my spirit to soar,*
*My **"Husband—Deepak Karwal"**, for helping me dream and giving strength to my wings,*
*My **"Children—Amaya and Pavit"** who, with their gentle applause and unwavering confidence give me the quiet courage to move forward.*

***—Dr. Sarabjeet Kaur Kochhar**, Associate Professor, Department of Computer Science, Indraprastha College for Women, University of Delhi, India*

In the loving memory of my parents!

***—Dr. Saibal K. Pal**, Scientist and Director, DRDO, India*

Preface

As artificial intelligence becomes deeply embedded in our digital ecosystem, the ability to make machines selectively "forget" has evolved from a theoretical curiosity into a fundamental necessity. "Machine Unlearning" offers the definitive guide to this rapidly advancing discipline that bridges privacy law, ethical AI development, and cutting-edge computational methods.

Machine unlearning—the targeted elimination of specific data influences from trained models—represents a paradigm shift in how we approach AI model governance. Driven by evolving privacy legislation including GDPR's "right to be forgotten" mandate and the urgent need to purge harmful, obsolete, or protected content from AI systems, this field addresses critical challenges that traditional model retraining cannot efficiently solve.

This authoritative book delivers a systematic exploration of machine unlearning's conceptual frameworks, algorithmic innovations, and real-world implementations across multiple industries. Starting with foundational principles and classification systems, readers master both exact unlearning methodologies—which guarantee complete data influence removal through sophisticated algorithmic approaches—and approximate strategies that achieve efficient influence reduction via targeted parameter modifications.

The text explores specialized unlearning protocols tailored to diverse computational architectures, spanning from transformer-based language models to graph neural networks and recommendation engines, while establishing robust methodologies for validating and measuring unlearning success. Industry case studies from healthcare informatics, financial services, personalization platforms, and autonomous systems illustrate how unlearning solves concrete challenges in regulatory compliance and data stewardship.

Beyond algorithmic considerations, this work addresses crucial ethical dimensions, security implications, and governance structures essential for responsible deployment. As AI systems penetrate increasingly sensitive applications, mastering selective information removal while preserving model performance becomes indispensable for practitioners, researchers, and policy architects.

Whether you're an AI engineer implementing privacy-preserving solutions, a researcher advancing theoretical understanding, or an executive, managing compliance obligations, this book delivers the authoritative knowledge required to deploy effective, efficient, and ethically sound machine unlearning in our privacy-first digital landscape.

Chapter 1: Introduction to Machine Unlearning—This chapter introduces the concept of machine unlearning, its importance in the current AI landscape, fundamental principles, and a comprehensive classification of techniques.

Chapter 2: Exact Unlearning Methods—This chapter delves into exact unlearning techniques that aim to completely remove the influence of specific data points from trained models.

Chapter 3: Approximate Unlearning Technique—This chapter explores approximate unlearning methods that efficiently minimize the influence of target data points through limited parameter updates.

Chapter 4: Machine Unlearning for Different Model Architectures—This chapter discusses how machine unlearning techniques can be applied to various machine learning model architectures.

Chapter 5: Verification and Evaluation of Unlearning—This chapter covers methods for verifying the effectiveness of machine unlearning and evaluating the performance of unlearned models.

Chapter 6: Applications of Machine Unlearning—This chapter presents real-world applications and use cases of machine unlearning across various domains.

Chapter 7: The Road Ahead: Ethical, Technical and Governance Perspectives for Holistic Machine Unlearning—This final chapter discusses the ethical implications, privacy and security aspects, governance issues, and future directions in the field of machine unlearning.

Author Contribution: Dr. Rajan Gupta has contributed to Chaps. 3, 5 and 6. Dr. Anamika Gupta has contributed to Chaps. 2 and 4. Dr. Sarabjeet Kaur has contributed to Chaps. 1 and 7. Dr. Saibal K. Pal has contributed to Chaps. 5 and 6.

Ciudad Victoria, Mexico — Rajan Gupta
New Delhi, India — Anamika Gupta
New Delhi, India — Sarabjeet Kaur Kochhar
New Delhi, India — Saibal K. Pal

Acknowledgments

The authors of this book would like to gratefully and sincerely thank all the people who have supported them during the journey of writing this book, to only some of whom it is possible to mention here.

Dr. Rajan Gupta would like to thank Prof. (Dr.) Fernando Rodriguez from Artificial Intelligence and Innovation (AI&I) Lab, Tamaulipas Autonomous University (UAT), Mexico, for his support and constant guidance towards new work in the field of Artificial Intelligence. Then, Dr. Gupta would like to thank his Ph.D. Supervisors—Prof. Sunil Kumar Muttoo and Dr. Saibal K. Pal, along with current and former faculty members of the Department of Computer Science, University of Delhi—Prof. Naveen Kumar, Prof. Vasudha Bhatnagar, Prof. Punam Bedi, Prof. Neelima Gupta, Mr. P. K. Hazra, and Ms. Vidya Kulkarni, for their valuable guidance and research directions in the field of Computer Science and Data Science. Also, Dr. Gupta would like to thank faculty members from Center of Information Technologies and Applied Mathematics, University of Nova Gorica, Slovenia, led by Prof. Tanja Urbancic, Prof. Irina, Prof. Nada and Ms. Tea for their valuable support. They all helped provide knowledge, infrastructure and resources related to Doctoral and Post-doctoral Research work, which was in Technology, Data Science and Public Information Systems. The doctoral as well post-doctoral research work helped in forming the basis for this book.

Prof. (Dr.) Anamika Gupta is deeply grateful to a whole lot of people who helped directly or indirectly in this endeavour. I am grateful to my parents who imbibed in me hard work and perseverance, my parents-in-law for their life long support, my husband for his tireless encouragement, my children for the joy and light in my life. I am grateful to my college Principal, Prof. Poonam Verma, for her belief in my potential. I am grateful to Prof. Naveen Kumar and Prof. Vasudha Bhatnagar for being my research supervisors, along with faculty members and colleagues of the Department of Computer Science, University of Delhi. I am grateful to all my colleagues at Shaheed Sukhdev College of Business Studies for creating an environment of discussions and progression. I am thankful to the publishing house for giving an opportunity to show my potential as an author.

I, ***Dr. Sarabjeet Kaur Kochhar***, would like to acknowledge my parents, for their unwavering love, confidence in my abilities, and quiet sacrifices that built my foundation; my husband—who is my deepest confidante, my anchor and a steady source of encouragement; my children, whose enthusiasm in my ventures and faith in my wings have lighted my path, bringing me joy and gently guiding me forward. My sister has remained a steady wellspring of comfort and added warmth to my path. My sincere thanks to my in-laws for their support throughout my journey. I would like to acknowledge my Ph.D. Supervisor, Prof. Vasudha Bhatnagar, Department of Computer Science, University of Delhi, for being my constant source of inspiration. Her values, hard work and sincerity continue to guide my moral compass. Whenever I sit to write, I find myself guided by her example. I am also grateful to Prof. Naveen Kumar and Prof. Neelima Gupta, Department of Computer Science, University of Delhi for their guidance and generous support throughout my academic journey. I would also like to place on record my gratitude for the Principal, Indraprastha College for Women, University of Delhi for her support and guidance.

Finally, this work would not have been possible without the invaluable support from the reviewers, editors and the entire publishing team of Springer Nature, esp. Thomas Ditzinger and Gowrishankar Ayyasamy. This book also recognizes incredible support from the book's endorsers, and the authors' guru, mentors, family, and friends. So, the authors would like to thank them all from the bottom of their hearts. The authors also acknowledge the use of AI technologies in shaping and refining certain sections of this book, each carefully reviewed with attention and care.

Contents

About the Authors

Dr. Rajan Gupta is an AI Professional with 15+ years of combined experience in AI/ML Product and Services Delivery, Analytical Research, Consulting, Training and Teaching in the field of Data Science and Computer Science. His core experience lies in embedding and operationalizing AI/ML into scalable products, with a focus on delivery, implementation, and technical excellence backed up by research-oriented approach. He created and worked for different AI/ML Center of Excellence, efficiently managed analytical product implementation and consulting teams in various domains like EdTech, HealthTech, Telecom, Retail and Manufacturing for Fortune 500 companies. He is a visiting Sr. Researcher at Artificial Intelligence and Innovation Lab, Tamaulipas Autonomous University (UAT), Mexico.

He has worked on various analytical consulting assignments on problems related to the areas of Digital Government, Healthcare, Education, Retail and Insurance. He has delivered lectures at the University of Delhi and IMT—Ghaziabad for various academic sessions in Computer Science, Data Science, Information Security and Management, and has taken invited lectures at IITs, IIMs, DTU, MDI, DU and IOCL. He has also conducted several 1:1 live mentoring and training sessions related to upskilling of analytical career, preparations for analytical certifications and technical knowledge on Analytics Project Lifecycle for various data science professionals from reputed organizations like Verizon, Nokia, Cyient, Tech Mahindra, TCS, Ericsson and Anthem.

He has done his Ph.D. in Information Systems and Analytics from the Department of Computer Science, University of Delhi, and Post-doc in Data Science and Data Modelling from the Center of Information Technologies and Applied Mathematics, University of Nova Gorica, Slovenia, Europe. He is triple masters in Analytics, Management and Computer Applications. He has authored more than 125 publications including 7 books and multiple research papers in the areas of Public Information Systems, Artificial Intelligence (AI), Machine Learning (ML), Data Science, Information Technology, and Management.

He is one of the few Certified Analytics Professional (CAP-INFORMS) around the world and is serving as CAP Ambassador in Asia Region. He is the first non-US member of the prestigious Analytics Certification Board (ACB) of INFORMS,

USA. He has also been accredited with 'Graduate Statistician' from the American Statistical Association (ASA). He is UGC NET-JRF qualified and holds a certificate in Consulting from Consultancy Development Centre (CDC), DSIR, Ministry of Science and Technology, Government of India.

He has received prestigious "Standout AI Thought Leader" at 3AI Zenith Awards 2024, 2025; AI and GenAI Transformation Leader of the Year 2025 at Bharat GenAI and Analytics Summit and Awards 2025; Future First AI Visionary Award 2025 at Data and AI Future First Awards; "AI Makers 100" which is Top 100 Most Influential AI and Analytics Leaders Award at 3AI GCC-X Summit 2023 and 2025; "AI Changemaker Leader" under 3AI ACME Awards at BEYOND 2023; and "40 Under 40 Data Scientists" award for 2022 by Analytics India Magazine at MLDS 2022.

His areas of interest include Artificial General Intelligence (AGI), Generative AI (GAI), EdgeAI, Metaverse, Algorithmic Government, Hyper-automation, Network Science, Data Science, E-Governance, Public Information Systems, and Information Security. He has contributed to the E-Governance Development Index report by the United Nations (EGDI-2020). He is a member of the reviewer panel of multiple international journals and conferences. He has also delivered a talk as a panelist on Data Science Application for E-Governance on an international forum sponsored by International Data Engineering and Science Association (IDEAS), USA, and conducted a Global Workshop on "Inclusion of Marginalized Communities" through Electronic Governance and Analytics at ICEGOV-2020 hosted by United Nations University, amongst many Corporate events, Panel Discussion, Enterprise Webinars, Faculty Development Programs, News Channel Debates, Podcasts and Research events, as an AI/ML and Data Science expert.

LinkedIn Profile: https://www.linkedin.com/in/rajan-gupta-cap/.

Dr. Anamika Gupta is a Professor of Computer Science in Shaheed Sukhdev College of Business Studies (SSCBS), University of Delhi. She has 25 years of teaching experience at the University of Delhi and 5 years of industry experience at various organisations. She is also holding the position of Program Coordinator for the PG Diploma in Cyber Security and Law course at SSCBS.

Her teaching and research areas include data analysis, machine learning, and deep learning techniques in the areas of Image Processing, Medical Data Processing, Natural Language Processing, Cyber Security etc.

She completed her Ph.D. in data mining from the Department of Computer Science, University of Delhi in 2013. She has published around 50 publications, including research articles in reputed journals, conference proceedings, book chapters, and magazines. She is a member of the editorial board of several journals and a member of the program committee of several international conferences. She is a reviewer for several journals of high repute.

She has delivered several lectures, invited talks, and has been a resource person in workshops at IIT Mumbai, DRDO, Delhi Technical University, Several colleges of University of Delhi, Amity University, IP University to name a few.

She has co-authored NCERT Information Practices textbooks and reviewed NCERT computer science textbooks. Dr. Anamika Gupta may be reached at anamikargupta@sscbsdu.ac.in.

Dr. Sarabjeet Kaur Kochhar is an Associate Professor in the Department of Computer Science at Indraprastha College for Women, University of Delhi. She holds a Ph.D. in Computer Science from the University of Delhi, specializing in Knowledge Discovery in Databases. A life member of the Computer Society of India, she brings over two decades of teaching experience in areas such as Data Mining, Databases, Computer System Architecture, Programming Languages, and Data Analysis and Visualization. Her research interests span Knowledge Discovery in Databases, Machine Learning, Data Analytics, Image Processing, Natural Language Processing, Generative AI, and Large Language Models.

Dr. Kochhar has a strong academic and research portfolio, with numerous publications in reputed journals and conferences, many of which are indexed in Scopus, ESCI, and SCIE. She has contributed extensively to academic literature, authoring book chapters for high quality international book volumes. She is also an author of various e-lessons for the Institute of Life Long Learning at the University of Delhi. Her work is recognized for its impact and quality, with contributions to cutting-edge topics in her field.

Beyond her research and teaching, Dr. Kochhar is actively involved in academic and professional engagements. She serves on the editorial boards of several journals and conferences, has reviewed numerous research papers, book chapters and has been invited as a keynote speaker, session chair, and resource person for workshops at esteemed institutions. Additionally, she has presented her research at prominent national and international conferences.

Dr. Saibal K. Pal is Director and a Senior Scientist with Defense Research and Development Organization, Government of India, for many years and has been awarded "Scientist of the Year" from the Government of India. He received his Ph.D. in Computer Science from the University of Delhi and is an Invited Faculty and Research Guide at several national institutions. His areas of interest are Information and Network Security, Computational Intelligence, Information Systems, and Electronic Governance. He has more than 250 publications in books, journals, and international conference proceedings. He has contributed to a number of significant projects and international collaborations and is a member of national advisory committees.

Chapter 1
Introduction to Machine Unlearning

Abstract Machine learning powers countless modern innovations by enabling systems to learn from data, detect patterns, and improve predictions over time—all without explicit programming. But its success hinges on vast, often sensitive datasets, raising critical concerns. This chapter explores how such concerns set the stage for machine unlearning, which aims to remove specific data from models, an increasingly vital task in today's data-driven world. It delves into the nuances of modular unlearning and navigates the evolving global privacy landscape, analyzing the major data protection laws across regions, from the EU to the US, India, China, Brazil, Japan and Australia, and how they reflect their unique sociopolitical contexts. The chapter also confronts technical challenges before machine unlearning: the complex architecture of ML models, non-linear data dependencies, computational intensity of data erasure, and the challenge of proving complete removal. Finally, the chapter underscores the pressing need for efficient, verifiable unlearning methods and introduces key metrics for evaluating their effectiveness.

1.1 What is Machine Unlearning

The term Machine Unlearning was coined by (Cao and Yang 2015), antithetical to the term Machine Learning to imply a systematized and complete dismantling of the influence of specific data points from a data model built by machine learning algorithms (Floridi 2023, Nguyen et al. 2022). Notably, the idea is not just to delete the data, unlearning should essentially update the model's parameters as if it had never seen the data to be removed in the first place (Floridi 2023).

1.1.1 Motivation for Machine Unlearning

The world is abuzz with the impact of machine learning models, as evident by their increasing number of successful applications. Machine learning (ML), a branch of

R. Gupta et al., *Machine Unlearning: Concepts, Techniques and Applications*,
Studies in Computational Intelligence 1270,
https://doi.org/10.1007/978-3-032-18447-4_1

artificial intelligence (AI) focuses on enabling computers to learn directly from data, identify patterns, and make predictions without being explicitly programmed. In fact, ML models can even improvise themselves over time by being exposed to more data. Naturally, thus arises the need for large amounts of data to train the ML models effectively.

The catch is that this data often includes private information such as the details related to users' identity such as their names, addresses, phone numbers, email addresses; Financial data such as bank account details, credit card numbers, transaction history; Health data such as medical records, diagnoses, treatments and Behavioral data such as browsing history, purchase history, app usage and social media activity, etc. (Bourtoule et al. 2021, Mercuri et al. 2022).

Infact private data is very useful for training ML models that endeavor to provide personalized recommendations such as recommending products, movies, or news articles based on user preferences. ML models employed for automated decision making, identifying potentially fraudulent transactions, assisting doctors in diagnosing diseases, and predicting future trends and events also depend heavily on the personal data for their efficacy.

The use of personal information for training ML models however leads to multiple problems. The most significant problems, that also set the motivation for machine unlearning (Fig. 1.1), are enumerated below:

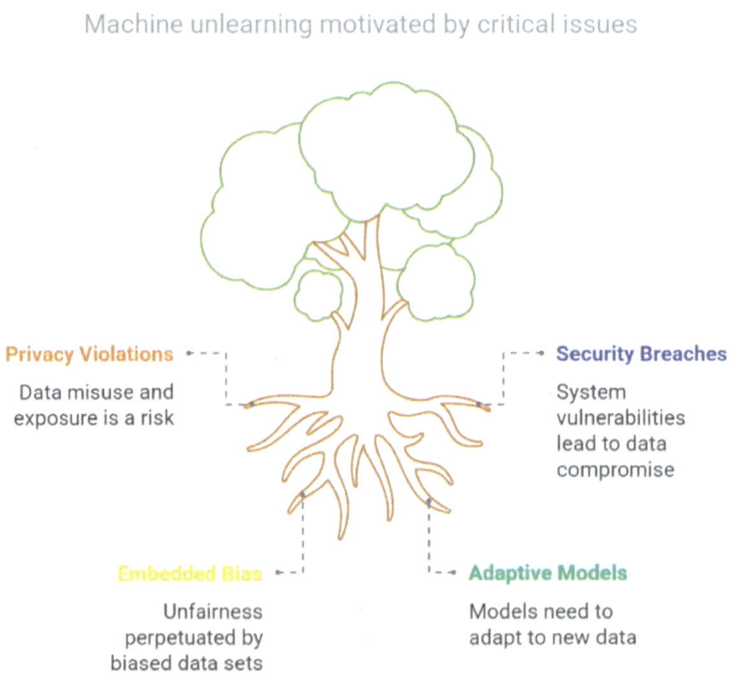

Fig. 1.1 The forces accelerating the need for machine unlearning

1. **Privacy Violations**: ML models can inadvertently lead to privacy infringements by revealing sensitive information about individuals, exposing them to risks such as unauthorized profiling and identity thefts. Even if private data is anonymized, techniques such as model inversion, reverse-engineering / model extraction, and targeted querying / membership inference can still enable links back to original records.

For example, model inversion attacks have been shown to reconstruct sensitive attributes from model outputs (Fredrikson et al. 2015). Model extraction or reverse-engineering via query APIs can reproduce model behavior and reveal training related information (Tramèr et al. 2016). Membership-inference and targeted-query attacks can determine whether specific records were present in the training set (Shokri et al. 2017). More recently, data-extraction work has demonstrated that large language models may leak verbatim training data under certain conditions (Carlini et al. 2021). These studies illustrate why weak or reversible anonymization and model overfitting create real privacy risks. The need to protect personal data is precisely why regulations like GDPR (Right to Be Forgotten) and CCPA (California Consumer Privacy Act) were enacted, requiring organizations to erase personal data upon request (further discussed in Sect. 1.4). Machine unlearning thus serves as a crucial tool to ensure complete data oblivion to protect an individual's privacy and adherence to the privacy regulations.

2. **Security Breaches**: Prone to cyberattacks and leaks, ML models can become involuntary facilitators of security breaches. The insecure systems naturally also open doorway to unauthorized personal data access. Typically, to orchestrate a security breach, the original training data is poisoned with fake data or noise is added in an adversarial attack, hence leading the same machine learning algorithms to generate a model with changed normative behaviour and erroneous predictions (Cao and Yang 2015, Nguyen et al. 2022). To restore system security, it is not sufficient to simply retrain or patch the model; instead, the poisoned or adversarial data must be precisely identified and completely unlearned from the trained model. This requirement naturally positions machine unlearning as a critical defence mechanism, enabling the removal of harmful data influences while preserving the utility of the remaining model. Recent advances in federated learning and distributed unlearning further emphasize this role, where compromised client updates or malicious local data can be selectively removed without retraining the entire system. By eliminating the residual effects of poisoned data, unlearning helps reestablish correct model behaviour and strengthens the overall security of learning systems.
3. **Embedded Bias and Ethical Concerns**: Personal data comes with an inherent bias (Shaik et al. 2024). If the training data is derived from sources that are prejudiced, distorted or skewed, the model built on it will not just immortalize such biases but may often amplify them. The outputs of such a model will therefore be non-transparent, and unfair towards a people of certain caste, creed, colour or sex. Machine unlearning is imperative to unlearn the discriminatory data points and improve fidelity of machine learning systems.

4. **The Need for Adaptive Models**: The data used for training a ML model may undergo a phenomenon called Concept Drift, wherein the data becomes outdated or irrelevant over time (Zhang et al. 2023). This occurs because real-world conditions are not static; for instance, users' preferences, their medical status and personal details may get updated. In healthcare settings, concept drift can arise when patient records are revised or withdrawn, requiring the removal of outdated or sensitive medical information from trained models. Similarly, in recommendation systems, previously learned user preferences may no longer reflect current interests and must be unlearned to avoid inaccurate suggestions.

Moreover, users can exercise their right to pull back their data even after a model has been trained on it. Machine unlearning algorithms allow unlearning of obsolete, unwanted and even erroneous or noisy data points from the original training data, allowing the ML model to adapt, stay accurate and efficient.

1.1.2 Machine Updation Versus Machine Unlearning

The real world applications in which machine learning systems are deployed continue to evolve. It is therefore critical to have the ability to modify machine learning systems, post their deployment, in order to retain the capability to control them and help them stay accurate and relevant. Machine Updation (or Model updation) and Machine unlearning serve as two mechanisms to accomplish this objective. It is necessary to distinguish between these seemingly similar terms to understand nuances of their fundamentally different ways of achieving the same objective and their different implications.

While machine updating involves enhancing the model by incorporating new data, machine unlearning is concerned with removing the influence of specific data points, also thereby ensuring compliance with privacy norms. In practical deployments, these mechanisms are often used together in hybrid workflows, where harmful, outdated, or non-compliant data is first unlearned, followed by model updating using clean and relevant data. Such a "first unlearn, then update" strategy is particularly useful for practitioners managing the full lifecycle of deployed models (Nguyen et al. 2022, Bourtoule et al. 2021). Table 1.1 presents the What (basic objective), When (used) and How (approach) of the two mechanisms and highlights other key differences between them.

1.1.3 Exact Versus Approximate Unlearning

Retraining a model from scratch, often referred to as naive retraining, may appear to be the most straightforward and reliable approach for achieving complete forgetting. By removing the data to be unlearned and training the model anew, one can

Table 1.1 Machine updation versus machine unlearning

Basis	Machine updation	Machine unlearning
What	ML model evolution due to change or addition of new information	ML model evolution due to removal of specific existing data points from a data model
When	Presence of a concept drift or availability of new data	Explicit data deletion request
How	Retraining or fine-tuning the ML model with new or changed data	Retraining whole or parts of model using special unlearning algorithms without the need for external data
Maturity	Well researched	Novel, under study
Complexity	Well managed	Usually complex
Legal compliance	Not generally required	Requires compliance to the data privacy laws
Evaluation criteria	Performance gains	Completeness of data removal
Examples	Updating a classifier based on new clinical insights or requiring better feature interpretation	Deleting a user's data upon receiving right to be forgotten request

theoretically ensure that no residual information from the deleted data remains in the final model. However, despite its conceptual simplicity, this approach is rarely practical in real-world settings. The computational cost of full retraining, especially for large-scale models and continuously evolving systems, makes it infeasible in terms of time, energy, and resource consumption (Mercuri et al. 2022).

As a result, recent research has increasingly focused on selective retraining, which aims to remove the influence of specific data points while preserving the model's overall performance. In this context, exact unlearning refers to methods that guarantee behaviour equivalent to full retraining, but only with respect to the data being removed. These methods strive for strong theoretical guarantees, ensuring that the unlearned model is indistinguishable from a model trained without the deleted data. While exact unlearning offers high fidelity, it often comes at the cost of increased algorithmic complexity and computational overhead.

In contrast, approximate unlearning methods trade strict guarantees for efficiency and scalability. Instead of fully eliminating all traces of the data, these approaches aim to reduce its influence to a negligible level, often using optimization shortcuts or parameter updates. Approximate unlearning is particularly attractive for large models and distributed systems, where exact retraining is prohibitively expensive. Although such methods may not achieve perfect forgetting, they provide a practical balance between computational efficiency and acceptable unlearning quality, which explains why they are widely adopted in modern machine unlearning research.

1.1.4 Modular Unlearning

Scalability, privacy preservation, and a clean modular design are desirable characteristics for the unlearning algorithms. Therefore, it is imperative to divide an unlearning process into units that are discrete, self-manageable, and can be independently modified or removed. Structured/Scalable Modular Unlearning (SAMUL) is a form of modular unlearning that breaks down the unlearning process into several organized layers, like a hierarchy. Layering allows for pinpointing and forgetting only specific, targeted pieces of information from earmarked parts of the system. Resultantly, SAMUL provides a scalable approach to unlearning, especially for unlearning in complex machine learning systems like distributed systems (which are spread across many different computers) and multi-functional systems (which handle a variety of tasks). Depending on the scope of what is being forgotten, the various levels of modular unlearning are summarized in Fig. 1.2 and can be enumerated as follows:

1. **Data Level Unlearning**

   This type of unlearning targets the removal of data at different granular levels, namely:

   Point wise Level: point-wise data level unlearning aims to remove the influence of specific and clearly defined subsets from within the training data points (records). The broader unlearning strategies (discussed later) are different in that they target entire features, classes, or modules. Hence, point-wise data-level unlearning symbolizes the most fine-grained information removal. For example, a user might request that the record of one or two particular items purchased in a transaction be removed.

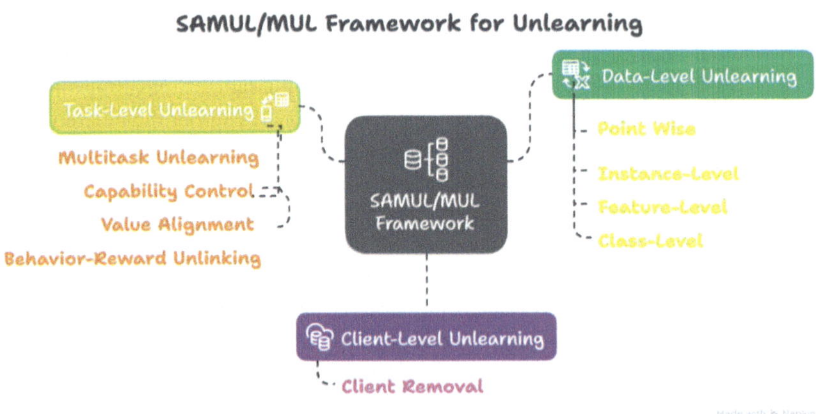

Fig. 1.2 Structured modular unlearning

Instance-Level: At this level, the focus is on the removal of specific data records associated with a specific real-world event, object, or entity, shifting focus to a higher-level conceptual "instance". This level of unlearning directly addresses the "right to be forgotten" type of removal requests and is particularly useful for privacy preservation settings. For example, if a user asks a cloud-based image background removal system to forget a face, instance-level unlearning would identify and remove all the instances of a person X's face from different images, rather than just one specific image file. It would also remove the impact of the original face from the model without disturbing its effectiveness to recognize other faces.

Feature-Level: Erases the contribution of a specific column or attribute or feature (e.g., age, gender, designation) from a machine learning model's influence. Feature-level unlearning is broader than instance-level unlearning, where the focus is just to erase the impact of a particular data records. Removal of an entire feature is particularly important in some settings where it is important to remove bias. For example, in a loan default model, while instance-level unlearning would allow erasure of the effect of some particular users' financial history (records), feature-level unlearning will allow removal of entire attributes such as gender or marital status from the training set, to address fairness concerns.

Class-level: Unlearns all data and model behavior associated with a specific category or class label. Class-level unlearning is broader than pointwise, instance-level and feature-level unlearning as it erases everything the model has learned about a specific class. Class-level unlearning becomes important in situations where it is important to save a particular group legally or ethically–for instance a group of students who are labelled'Slow Learners' or a group of people with a sensitive medical condition, say,'COVID Positive'.

2. **Client-Level Unlearning**

   This type of learning operates at a scope higher than the data-level unlearning. Client-level unlearning removes all the data or contribution (records, updates, and gradients) associated with a particular client (user or device). Client-level unlearning works with the federated database learning systems, which are essentially machine learning systems that consist of multiple devices or local clients like smartphones or sensors etc. A central server maintains a global ML model. All the clients train their local ML model on their private data and send back their updated model parameters (like weights or gradients) to the central server. In response, the central server aggregates these to update the global model. This process is generally re-iterated to improve the global central model.

3. **Task-Level Unlearning**

   This sort of unlearning pertains to multi-task models and entails erasing a complete learned task, typically without diminishing its effectiveness on other tasks. In essence, therefore, task-level unlearning represents selective forgetting.

For instance, forgetting how to cluster particular type of objects. Let us discuss some important task-level unlearning components:

Multitask Unlearning: serves as a core mechanism for task-level unlearning, with the goal of entirely eliminating the model's capability to execute a particular task it was intentionally trained for. The result is that the model fully forgets how to perform the targeted task. Multi-task unlearning also involves removal of the knowledge and parameters of the said task from the model.

Capability Control Systems: are a kind of behavioral restriction imposed on the task-level unlearning systems. Unlike multitask unlearning systems, they don't unlearn how to do a task or delete its internal knowledge and weights etc. They simply control the execution of the task by restricting or disabling the model's capability to use its knowledge, generally for adherence to a policy, or for safety, or ethical reasons.

Value Alignment: is a design objective deployed after unlearning a specific task to remove any residual behaviors. The intent is to align the remaining tasks and responses of the unlearned model with the current goals and ensure that the forbidden knowledge isn't accidentally revived through related functions. Such an alignment of a model's behavior is largely for conformance with ethical, social or organizational guidelines.

Behaviour–Reward Unlinking: is a supporting mechanism for task unlearning, used to remove a learned behaviour or a skill that was acquired through reward-driven learning. Reward-driven learning is generally used in reinforcement learning, a type of machine learning technique where, instead of learning from correct answers, learning is achieved by associating a task with rewards and punishments, based on actions performed. The behaviour–reward unlinking works by erasing the learned mappings between actions and rewards, thereby removing the motivation and memory to perform the action. For example, in a video game supported by reinforcement learning, a robot is trained to jump over obstacles to earn reward points. To unlearn jumping, behavior-reward unlinking removes the credit of reward points with the action'jump'.

A comprehensive benchmark that unifies deletion sets and trained models across multiple tasks and data modalities, including speech and video classification (Cheng, J., & Amiri, H., 2024). It enables consistent evaluation of unlearning performance and scalability for general unlearning algorithms across diverse tasks. MUSE (Huang et al., 2024) focuses on evaluating unlearning for large language models using six comprehensive criteria — including memorization, privacy leakage, utility preservation, and scalability of sequential unlearning. This benchmark is particularly relevant for task-level and data-level unlearning in models trained on large text corpora.

Despite its advantages, modular machine unlearning introduces several technical challenges that must be carefully managed. One major concern is catastrophic forgetting, where the removal of specific data or tasks unintentionally degrades the model's performance on retained knowledge. There is also a fundamental trade-off between speed and fidelity, as faster unlearning methods may only approximate true forgetting, while exact unlearning can be computationally expensive. In distributed and federated settings, additional complexity arises during model re-aggregation, where selectively unlearned updates must be safely merged back into a global model without reintroducing removed information or harming overall consistency.

1.2 Why isn't It Simple to Forget–Challenges Before Machine Unlearning

As discussed in the previous section, in view of the ever-growing demand for user control, privacy frameworks, and model evolution, machine unlearning is gaining prominence by each passing hour. However, the path of complete and efficient data forgetting is dotted with a host of practical and technical obstacles that stem from the way the machine learning models are built, their non-linear dependencies on data, the high computational costs for complete and exact removal of data, a required guarantee of data removal, and high proficiency of the removal method. In the following discussion, we shed some light on these issues that pose challenges to the effective growth of technology (Fig. 1.3).

Complete Unlearning

It implies removing the data from the model as if the model were never exposed to those data during training. Though desirable, complete unlearning of data is not always practically possible. An ML model is built on insights from the complete set of available data points. The data point itself can be removed from the input data but it is very difficult to understand, track and remove the influence of that data point on the model and its learned parameters. Hence, even after removal, the model may indirectly remember the data via its influence.

Exact Unlearning

It involves deleting the exact data point in question. However, as discussed above, an ML model's knowledge, patterns, and dependencies are a function of the training data as a whole, and not separate data points. Consequently, the effect of a single data point can be scattered across the model numerous, incorporated amongst its various parameters, making it difficult to isolate and eliminate.

Box 1: Complete unlearning versus Exact unlearning

Complete unlearning means that all influence of a particular data point or dataset is removed from the model, so the model's behaviour is indistinguishable from one that never used that data, even if the internal parameters are

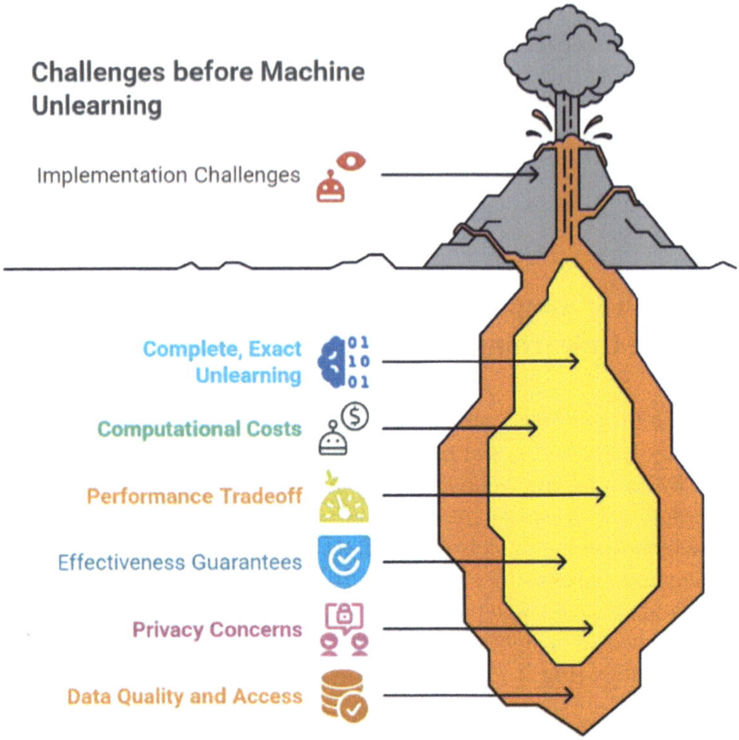

Fig. 1.3 Machine unlearning: hurdles to overcome

different. **Exact unlearning**, on the other hand, requires a stronger condition: after removing the data, the resulting model must be *identical* to the model that would have been obtained by retraining from scratch on the remaining data, accounting for all training randomness.
So, exact unlearning is a strict, algorithmic guarantee; complete unlearning is a semantic, behavioural guarantee. Exact unlearning $\Rightarrow$ complete unlearning, but complete unlearning $\not\Rightarrow$ exact unlearning.
Example: Suppose a sentiment analysis model is trained on one million reviews, and one user asks for their review to be deleted. With exact unlearning, the updated model is exactly the same as a model retrained from scratch on the other 999,999 reviews. With complete unlearning, the model no longer memorizes, leaks, or depends on that user's review in any measurable way, but its internal weights may still differ slightly from a fully retrained model.

Computational Costs

The twin requirements of complete and exact unlearning often lead to the most straightforward verifiable way of unlearning–to retrain the model from scratch on the data excluding the data to be forgotten–called as Naive Retraining (Sect. 1.1). However, for real-world applications that operate on large datasets and complex models, Naive Retraining becomes impractical and unscalable due to its prohibitive computational and time-consuming requirements.

The Performance Trade-Off

The relationships learned by an ML model from data points may be complex and non-linear. As noted in the points above, removal of selected data points is not just very difficult and exorbitantly costly, but disrupt the data inter-dependencies, severing patterns and insights, possibly downgrading the model's overall performance, accuracy and generalizability. Hence a delicate balance needs to be maintained between completely forgetting the data and the performance of machine. For instance let us consider deep learning models which embed data deeply and irreversibly. Deep learning models are trained using Gradient Descent–a process in which the weights are incrementally updated to predict error. There's no way to trace how a specific data point influenced the final model weights and reverse these updates to remove the effect of one data point. Similarly, application of different sets of weights in a deep learning network could result in similar performance (called non-convex optimization). Removal of a data point from such a space may lead to a model that may still behave similar to the original model or produce a completely different model that is not at all identical to the original model. For deep learning models therefore, costly naive retraining needs to be performed, sacrificing the performance of model to achieve complete unlearning.

Effectiveness and the Unlearning Guarantees

The regulatory privacy frameworks can be very stringent, demanding a guarantee of complete erasure of data, especially in domains like healthcare and finance. However, it is pertinent to be able to access and compare with the original data to evaluate the effectiveness of the updated model and certify guaranteed data removal. Since original data access is restricted, churning out verifiable guarantees are a significant challenge. Moreover, the development of metrics for evaluating the efficacy of unlearning is still an active research area, brimming with challenges. The various directions in which these research endeavors are focused are discussed in Sect. 1.3.

Privacy

There is a need for privacy protection even during the unlearning process. Subtle clues like the time required to unlearn a data point or shifts in model behavior may lead to potential exposure of information related to the data meant to be erased. Hence as noted in point 5, the access to the original data has to be restricted to protect data privacy and data access can't even be granted for establishing the effectiveness of the machine unlearning process.

Data Quality and Access

As is the case with traditional machine learning algorithms, the quality of machine unlearning also suffers from the GIGO (Garbage in Garbage out) mode. For instance, noisy (incorrect) labels make it difficult to exactly identify the specific data points that need to be unlearned; Weak supervision (inaccurate, incomplete, or imprecise labels matched with input features) makes it difficult to guide the unlearning algorithm when labels are unreliable and restricted access to original data makes it difficult to verify that the unlearning process has been successful.

In essence, the challenges stem from the opaque, and intertwined way in which machine learning models encode knowledge, rendering the secure and verifiable extraction of particular information a complex task. Challenges such as catastrophic forgetting, verification of forgetting, scalability to large models, and deployment in distributed or federated settings further complicate practical adoption. These difficulties highlight the gap between theoretical unlearning guarantees and reliable real-world implementations. These challenges underscore the need for more efficient, robust, and provably effective machine unlearning techniques that can address the growing demands for data privacy in the age of pervasive machine learning. This challenge is succinctly captured in the following observation from a widely cited survey:

> "The complex web of dependencies learned by modern machine learning models makes holistic, verifiable forgetting profoundly hard."— adapted from Nguyen et al., 2022, ACM Computing Surveys

This perspective underscores why the next step in machine unlearning research focuses not only on algorithms, but also on metrics and benchmarks that can meaningfully measure forgetting quality and residual information leakage. The next section discusses metrics that evaluate the efficacy of machine learning algorithms.

1.3 How Well Does an Unlearning Algorithm Delete the Requested Data?

Evaluation phase is as equally important a phase in the life cycle of every machine unlearning algorithm, as are its development and implementation phases. The objectives of the evaluation phase are multi fold: the primary objective definitely being to establish the complete and precise forgetting of the data in question. These primary goals are in alignment with the requirement of privacy regulations, outlined in the next Section. The secondary goals of effective evaluation include establishing that the algorithm is efficient and privacy-preserving. As we speak, research for new measures and metrics to assess the effectiveness of unlearning is underway. This section outlines major pathways for developing evaluation metrics and mentions some specific measures and metrics pertaining to each pathway. We delve into topics such as residual influence analysis, comparisons with retrained models, resistance to privacy threats, and maintaining performance. A thorough exploration of the specific characteristics and formal mathematical definitions will be addressed comprehensively in a separate chapter.

1.3.1 Residual Influence Metrics

Residual Influence refers to a set of evaluation metrics that determine the precision of unlearning by measuring the influence that the deleted data still has on the ML model after unlearning. Influence functions are foundational in the most cited machine unlearning literature because they estimate how removing a single training point would change a model's parameters or predictions, enabling both auditing and approximate unlearning. This idea was popularized in the widely cited work (Koh and Liang, ICML 2017), which connects classical robust statistics to modern deep learning models.

Approach: Compare the unlearned model to a model "retrained" from scratch on the original dataset excluding the data points to be forgotten.

Goal: Perfect unlearning–The residual influence should be zero or near-zero i.e. If the unlearned model's parameters or behavior are identical or very close to the retrained model, it signifies the model behaves as if it had never seen the target data.

Metrics: Some commonly used residual influence metrics are listed below:

Prediction Difference (ΔPred)

Used generally in simple classification models, it measures change in model's output before and after unlearning. The general format of the prediction difference metric is:

$$\Delta Pred = Pred(Retrained\ Model) - Pred\ (Original(unlearned)\ Model)$$

Log Loss Change

Also called Cross-Entropy Loss, it is essentially a loss function that checks how well a model predicts the true label of a sample, given its predicted probability distribution. Its general form may be written as:

$$LogLoss = \log\,(probability\ of\ TRUE\ class)$$

To measure residential influence, the deviation in log loss may be computed as follows:

$$\begin{aligned}\Delta LogLoss &= LogLoss(Retrained\ Model)\\ &\quad - LogLoss(Original(unlearned)Model)\end{aligned}$$

A positive, higher value of the ΔLogLoss will indicate that the prediction has become less confident or incorrect implying that the model truly "forgotten" the point. Both the prediction difference and the log loss metrics capture the difference in the outputs of the original and retrained models. However, the prediction difference is more generic. It can consider all class probabilities or output changes, whereas log loss function measures how well the model predicts the correct class (confidence) and therefore focuses on the predicted probability for the true label only.

Influence Score

This is a class of metrics that try to measure the influence of the data point to be removed (say p_i), while predicting a new test point (say p_{test}). A generalized formula for measuring the influence can be stated as:

Influence (p_i, p_{test}) = (How Sensitive the Test Point is) \Times (How Important the Training Point is)

Where:

Sensitivity of the test point is measured by ascertaining how much the test prediction depends on the model's parameters and Importance of a training point is measured by gauging how much the training point affects the model's parameters.

Interpretation:

If the influence is nearly zero, the model probably forgot it, otherwise the unlearning was ineffective i.e.

- If Influence(p_i, p_{test}) is near zero → the removal of point p_i will not have an impact on predicting $p_{test.}$
- If Influence(p_i, p_{test}) is large and positive → removing p_i would negatively impact prediction of $p_{test.}$
- If Influence(p_i, p_{test}) is large and negative → removing p_i might even improve the prediction of $p_{test.}$

Attribution Function

This is a class of metrics that checks if the forgotten data still affects the model by tracking the points which are responsible for particular predictions. The different methods that fall under the umbrella of attribution function metric, however answer the aforementioned question using different mathematical approaches and practical considerations. For instance, TracIn is better suited to deep learning models as it uses gradient tracking during training. SHAP (Shapley values) uses game-theory to track influence of the training points. Methods like LIME and Grad-CAM scale better to large datasets than the computationally intensive SHAP and offer better interpretability and speed. SHAP however offers better accuracy. Thus, different methods present a distinct balance among accuracy, effectiveness, and adaptability. It is also important to note that while methods such as SHAP and LIME are valuable for inference and interpretability, attribution alone does not guarantee regulatory compliance or data deletion guarantees; however, these techniques are essential for internal model auditing, diagnostics, and understanding how individual data points or features influence model behaviour.

1.3.2 Consistency with Retrained Model

Consistency with Retrained Model refers to a set of evaluation metrics to determine the similarity (or dissimilarity) of the unlearned model to a model that was retrained from scratch without the deleted data. In real-world deployments, fully retraining a model for every data deletion request is often computationally infeasible, so practitioners rely on approximate measures—such as parameter, representation, or prediction similarity—to assess unlearning effectiveness instead of exhaustive retrain-and-compare procedures.

Approach: Like Residual Influence, this model also compares the unlearned model to a model "retrained" from scratch on the original dataset excluding the data points to be forgotten, but unlike the Residual Influence model that measures the influence of deleted data, this model computes the convergence or divergence of the two models.

Goal: The more similar the outputs and behavior of the unlearned model and the retrained model are, the more effective the unlearning.

Metrics: Prediction Difference and Log Loss change metrics, discussed in the previous subsection are also sometimes used to measure divergence. Some other metrics that fall under the umbrella of Consistency with Retrained Model, namely, KL divergence, Total Variation and Cosine similarity are discussed below:

Kullback–Leibler (KL) Divergence

This metric is used for probabilistic models and measures the divergence between the output probability distributions the retrained model, say, P(x) (also called True/ Reference distribution) and the unlearned model say Q(x) (also called Approximate/ Current distribution). General formula for KL Divergence is:

$$KL_D = \int_{-\infty}^{\infty} P(x) \log\left(\frac{P(x)}{Q(x)}\right) dx$$

Interpretation: KLD would add up how often the unlearned model was wrong with respect to the retrained one and how badly. A near zero value would mean that the two distributions are very similar whereas a bigger number would imply more difference.

Total Variation Distance

This is a symmetric measure, bound between 0 and 1, and is stricter than KL Divergence. While KL Divergence measures difference between two distributions, Total Variation measures the maximum possible difference in probabilities over all outcomes.

Cosine Similarity

Let A$^{\rightarrow}$ and B$^{\rightarrow}$ be two vectors representing the feature embeddings/ predicted probability distributions/ hidden representations or simply stated the outputs of the untrained and retrained models. The cosine similarity between the two can be measured as per the following mathematical formula:

$$Cosine\,Similarity = \frac{\vec{A} \cdot \vec{B}}{\left|\vec{A}\right|\left|\vec{B}\right|}$$

Interpretation: The metric essentially computes the fraction of (How much the vectors point in the same direction) ÷ (Magnitude of the vectors).

If Cosine Similarity = 1.0, the two vectors are identical in direction implying perfectly consistent predictions.

If Cosine Similarity = 0, the two vectors are are orthogonal in direction, implying no similarity.

If Cosine Similarity = –1.0, the two vectors point in opposite directions, implying completely opposite predictions.

Unlike KL Divergence and Total Variation Distance that measure dissimilarity of probability distributions, Cosine similarity measures similarity of the direction of vectors and has a range of [–1,1].

1.3.3 Privacy Leakage/Membership Inference Resistance

Privacy Leakage in the context of machine unlearning refers to the risk that an attacker can find some clues about the identity of the already deleted data in the unlearned model.

Approach: Generally, a Membership Inference Attack (MIA) is launched on the unlearned system to determine beforehand whether an attacker will be able to ascertain that a particular data point was part of a model's training set. Recent high-profile results show that strong machine unlearning should render previously seen ("in") and unseen ("out") data statistically indistinguishable to an attacker. In such cases, the accuracy of Membership Inference Attacks (MIAs) drops close to random chance, indicating minimal privacy leakage from the unlearned model. As a result, evaluating privacy leakage via MIAs has become a standard criterion for validating unlearning effectiveness and is increasingly required in regulatory audits, particularly in sensitive domains such as healthcare and finance.

Goal: To build a model that is capable of withstanding MIA, referred to as Membership Inference Resistance. Models that show differing predictions for data from the training set versus new data can reveal previously seen data and are known as unresistant. Conversely, a resistant model maintains consistent predictions regardless of the data point's presence in the training set.

Metrics: Some commonly used metrics are listed below:

Attack Accuracy

This metric measures the accuracy of a Membership Inference Attack. It is implemented as a binary classifier that tries to predict whether a data point was member of the training set (binary 1) or was a non member (binary 0). The general formula for attack accuracy can be listed as under:

$$Accuracy_{\{Attack\}} = \frac{Number\ of\ Correct\ Predictions}{Total\ Number\ of\ Predictions}$$

Interpretation:

If $\text{Accuracy}_{\text{Attack}} = 50\%$, attacker doesn't have an edge. Since the chances are 50% he cannot really tell if a point was in the training set.

If $\text{Accuracy}_{\text{Attack}} > 50\%$, the attacker has a better chance of knowing about the deleted data so there is a risk of privacy leak.

If $\text{Accuracy}_{\text{Attack}} < 50$, the attacker is doing worse than random guessing.

Precision/Recall/F1-Score

To launch a Membership Inference Attack (MIA), the system usually prepares an attack database. If this database is imbalanced, i.e. if there are more data points in the attack dataset which did not belong to the training set (non-members) than the members, attack accuracy metric may become misleading. For instance, if the MIA predicts" non-member", it will contribute to accuracy, but this accuracy of MIA is not actually true. In case of imbalance attack database therefore metrics such as Precision, Recall and F1-Score are used.

Let TP denote the term True Positives (i.e. correctly predicted"members"), FP denote False Positives (predicted"member" but actually"non-member"), FN denote False Negatives (predicted"non-member" but actually"member") and TN denote True Negatives (correctly predicted"non-members"). These terms can be better understood by looking at the Confusion Matrix depicted in Fig. 1.4.

Predicted Value / Actual Value	Member	Non Member
Member	True Positive (TP)	False Negative (FN)
Non Member	False Positive (FP)	True Negative (TN)

Fig. 1.4 Confusion matrix

Precision: is used to compute how many of predicted members were actually members, using the following formula:

$$Precision = \frac{TP}{TP + FP}$$

Recall: is used to compute how many actual members were correctly identified.

$$Recall = \frac{TP}{TP + FP}$$

F1-score: computes the balance between precision and recall and is computed by finding their harmonic mean:

$$F1 = 2 \times \frac{Precision \times Recall}{Precision + Recall}$$

1.3.4 Efficiency Metrics

Efficiency Metrics measure the speed and resources (memory and computational cost) used by an unlearning algorithm. They are also used to set benchmarks for unlearning algorithms.

Approach: Measure the time to unlearn, the amount of memory required for unlearning and compare them with the time and resources required for full retraining. Modern machine unlearning frameworks explicitly emphasize these efficiency metrics, with methods such as SISA and more recent Fairness-aware Machine Unlearning (FMU) reporting benchmark results on time, memory, and computational savings in published evaluations and leaderboards. For example, the FMU literature reports achieving unlearning costs of under 20% of full retraining cost while maintaining fairness and privacy guarantees, highlighting that effective unlearning can be substantially cheaper than retraining from scratch.

Goal: Minimize the time and computational cost of an unlearning algorithm, without compromising its correctness or privacy so that the cost of unlearning is cheaper than retraining.

Metrics: Some metrics used to determine efficiency of the unlearned model are:

Unlearning Time

This is the time to delete a data point. Lower the unlearning time, more efficient the unlearning algorithm. The general formula for measuring unlearning time is straightforward:

$$Time_{Unlearning} = Time_{End\ of\ Algorithm} - Time_{Start\ of\ Algorithm}$$

Memory Consumption

This is the RAM/VRAM required for running the unlearning algorithm. Like the unlearning time, less consumption of memory resources signals better algorithmic efficiency.

Retraining Effort

This measures the overhead of completely retraining an unlearning model instead of unlearning it. It is measured by the following fraction:

$$Retraining\ Cost = \frac{\text{Cost}\ of\ Unlearning}{\text{Cost}\ of\ Retraining}$$

Interpretation:

If the ratio $< 1 \rightarrow$ Unlearning is cheaper than retraining (desirable)
If the ratio $= 1 \rightarrow$ Unlearning offers no significant cost advantage
If the ratio $> 1 \rightarrow$ Unlearning is more expensive, which defeats its purpose

1.3.5 Performance Retention

Performance Retention metrics measure the usefulness, accuracy and effectiveness of how well a machine learning model performs typical machine learning tasks after unlearning.

Approach: Compute suitable metrics to evaluate the performance of the model post-unlearning, thereby facilitating a comparison with the original model.

Goal: Sustain performance while facilitating efficient unlearning of data.

Metrics: Metrics such as Accuracy, Precision, Recall, F1-Score (discussed under 3.3) and Log Loss (discussed under 3.1) are used to evaluate performance retention of the unlearned model. In safety–critical and regulated deployments, performance retention is monitored particularly closely. In practice, a small and well-quantified drop in accuracy is often acceptable if it enables significantly stronger privacy guarantees or more reliable forgetting, especially in domains such as healthcare, finance, and autonomous systems where data deletion obligations are strict. Recent machine unlearning studies therefore recommend reporting utility–privacy trade-off curves, which visualize model performance against privacy or forgetting strength, rather than relying on a single post-unlearning accuracy number. Such plots, now common in experimental sections of unlearning papers, make explicit the cost of stronger unlearning in terms of predictive utility.

Please also note that evaluation itself remains an active research area. Open challenges include how to reliably assess performance retention and forgetting in federated learning and large language model settings, where retraining baselines, data boundaries, and memorization behaviors are harder to define.

1.4 Privacy Regulations

As evident from the discussion in this chapter, the world is being increasingly influenced by data and the regulation of personal information has become a global priority. In such a scenario, it is important to understand the global landscape of privacy regulations set up by nations and the balance of challenges such as achieving interoperability and collaborative research across international borders. In this section, we explore the data protection frameworks of several key nations–including Europe, the United States, India, China, Brazil, Japan, and Australia–that have enacted comprehensive data privacy laws according to their own social, political, economic and legal systems (Fig. 1.5).

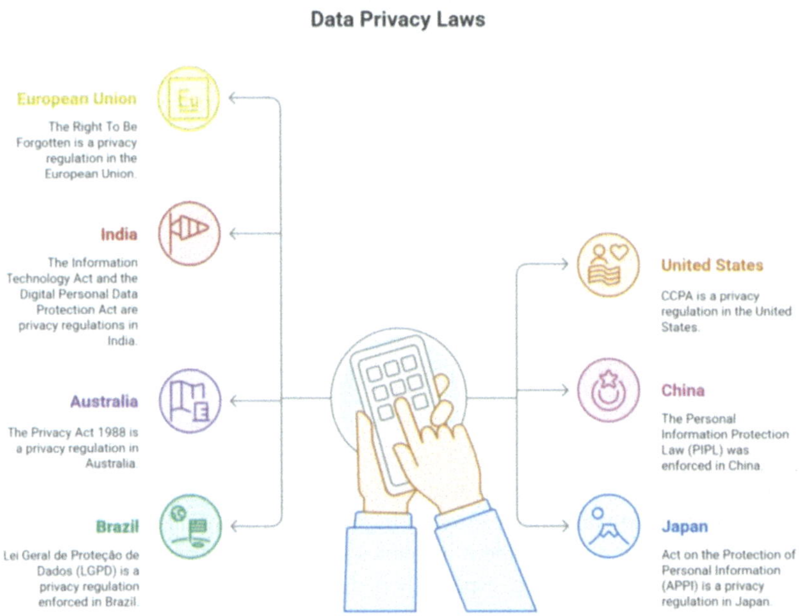

Fig. 1.5 Data Privacy Laws (Image Description: Diagram showing privacy regulations in European Union, India, Australia, Brazil, United States, China, and Japan)

1.4.1 European Union–the Right to Be Forgotten

The Right To Be Forgotten (RTBF), also known as the Right to Erasure, is a legal mechanism that empowers individuals to control the usage of their data on the digital platforms and is included in article 17 of the General Data Protection Regulation (GDPR). European Union drafted and enacted the GDPR to harmonize laws for data protection and privacy. GDPR came into effect in 2016 after it was passed by the European parliament and made obligatory for all organizations to comply vide May 25th, 2018 (Parliament and of the European Union 2018). The regulation formally defines the terms such as personal data, processing of data, restrictions on data processing, profiling, pseudonymization, and personal data breach etc. GDPR, under the Article 17 enshrines the Right to Erasure i.e. the Right To Be Forgotten, that mandates that an individual's personal data be completely erased when any of the following conditions are met:

- When the data is no longer being used for the original purpose that it was collected for;
- When the user withdraws her consent
- When the user objects to processing of her data
- When user's data has been processed in an unlawful manner
- When the data erasure is mandated by law
- When data is used for services that are specifically targeted for or are accessible to children for example social media platforms, online games and apps etc.

GDPR—particularly the Right to Be Forgotten—has been a primary driver behind the rapid evolution and growing academic focus on machine unlearning algorithms, as it created a concrete legal obligation to ensure that individuals' data can be effectively removed not only from databases but also from trained machine learning models. The Right to Be Forgotten (RTBF) strengthens the technical requirements for machine unlearning by extending deletion obligations beyond databases to trained models themselves. In practice, this implies a duty to ensure that predictive models, representations, and learned parameters no longer retain or exploit an individual's data, aligning legal compliance with formal unlearning guarantees rather than mere data record removal. Infact, a recent clarification by the European Court of Justice has reaffirmed the Right to Be Forgotten by upholding individuals' erasure rights and clarifying that online search engines—and, by extension, AI systems that surface or regenerate personal data—must meaningfully limit continued access to deleted personal information. There is a provision for imposition of very heavy penalties for violations to the standards documented in GDPR, signifying the firm stance of Europe on data privacy and security.

1.4.2 United States–CCPA

The California Consumer Privacy Act (CCPA) entitles the residents of California to a greater control over their personal information collected by commercial or other organizations (State of California, 2020). Under the CCPA, businesses are bound to explain privacy rights to their consumers and respond to their requests within a reasonable period of time. The California Privacy Protection Agency (CPPA) was established under the California Privacy Rights Act of 2020, on August 14, 2020, to implement CCPA. Under the CCPA, consumers have been granted the following rights to safeguard their privacy:

The Right to Know: The consumers can demand to know how their personal personal information is being used and shared by a business.

The Right to Delete: The consumers can insist the deletion of the personal information collected from them.

The Right to Opt-out: These right grants the consumers right not to share or sell their personal information.

The Right to Non-discrimination: This right entitles consumers who exercised their CCPA to not receive discriminatory treatment from businesses. The right prohibits businesses to deny goods or services, suggest, or charge different prices or quality for goods.

The CCPA was amended on January 1, 2023 to add the following two new privacy protections:

The Right to Correct: The consumers can get their personal data corrected.

The Right to Limit: Consumers can actively control the way sensitive personal information is used and disclosed.

While the CCPA's erasure and deletion requirements are generally narrower in scope than those of the GDPR, the law has become an influential template for privacy regulation across the United States, informing newer state-level frameworks such as the CPRA in California and the VCDPA in Virginia. Importantly for global readers, unlike the European Union, the U.S. still lacks a comprehensive federal privacy law, resulting in a fragmented regulatory landscape where obligations for data deletion and model unlearning vary significantly by state.

1.4.3 India–the Information Technology Act and the Digital Personal Data Protection Act

India is one of the largest data markets in the world. Although the Information Technology Act, 2000 (IT Act) was legislated by the Indian Parliament as early as October 2000, it did not address privacy, data protection, free speech, social media, and internet-related issues (Blythe 2006). An amendment to the IT Act, 2000, namely, the Information Technology (Amendment) Bill 2006 was enforced in October 2009 that included provisions related to individual privacy and data protection. Under

Section 69B of the amended act, rules such as the Information Technology (Procedure and Safeguards for Interception, Monitoring and decryption of Information) Rules, 2009, allowed the Government to intercept, monitor and decrypt information. Information Technology (Procedure and Safeguards for Blocking for Access of Information by Public) Rules, 2009 allowed blocking of websites for cyber security. The Information Technology (Intermediary Guidelines and Digital Media Ethics Code) Rules, 2021, formally notified on February 25, 2021, directed firms behind social media platforms, called intermediaries, to address user grievances and block, delete, or modify published content within strict time deadlines of 24 or 48 or 72 h and divulge the source of objectionable content upon court or government's directions.

Digital Personal Data Protection Act 2023 was enacted in India in 2023 in accordance with the declaration of "Privacy rights' as a fundamental right by the supreme court of India in 2017 (Manjunathan 2022). The bill, influenced by the European General Data Protection Right (GDPR), set up a Data Protection Authority (DPA) to protect the personal data of its citizens (Bareh 2024, Chopra and Mansharamani 2021). The bill is applicable to all the digital data generated within India or abroad, that is being used to provide some goods or services in India (Sengar 2023). The act primarily allows data principals (subjects of the data) to seek access to their digital data and the details of how it is being processed by the data fiduciaries. A data fiduciary is essentially a person or company that collects and stores the data in order to process it and use it for a specific purpose. Data principals can request the identities of the data fiduciaries and the other parties (if any), with whom the data has been shared. If the data principal is unhappy with the way their data is being used, she may request for data modification or file a grievance for inappropriate usage of data. If the data has outlived the purpose for which it was collected or the data subject simply wishes to withdraw the consent for the usage of her data, she may request even complete erasure. The patrons also have a right to nominate another person, in case of their mind or body incapacitation or death, to handle their personal data.

The Digital Personal Data Protection Act mandates data fiduciaries to implement the latest security protocols to protect personal data of the patrons and report any breach of data to both the Data Protection Board and the affected patrons. A Data Protection Officer (DPO), resident in India, must be appointed to deal with user's grievances and conduct periodic data compliance assessments with the help of an independent data auditor. Heavy penalties or the order of INR 10,000–INR 250 crore can be imposed by the Data Protection Board of India, depending on the nature and type of offences.

India's Digital Personal Data Protection (DPDP) framework explicitly strengthens the relevance of machine unlearning by granting citizens a right to request erasure that extends beyond raw databases to all downstream uses of personal data, including trained models and analytic outputs derived from it. Inspired in part by the GDPR, the Indian approach is still evolving in terms of regulatory enforcement and technical interpretation, and achieving international interoperability—especially for globally deployed AI models trained across jurisdictions—remains a significant and open challenge.

1.4.4 Data Privacy Laws in Other Countries

Countries such as China, Australia, Japan, and Brazil also have established comprehensive privacy laws to regulate the collection, use, and protection of personal data, each reflecting their unique legal and cultural contexts.

The Personal Information Protection Law (PIPL) was enforced in China, from November 1, 2021, by the Cyberspace Administration of China. The PIPL mandates that explicit consent be taken from data subjects before collecting, transferring or processing personal data, data related to health, finance and data of the children under the age of 14. The data subjects are also granted rights to access their data, get it updated or deleted, and to withdraw consent at any time. Moreover, Data Security Law (DSL), enacted in China on September 1, 2021, classifies data into three different categories, namely, the important data (the data that has potential to impact economic or social development of country), core data (the data that could impact national security) and the general data. The DSL implements different data security measures for the three categories of data. Regular risk assessments are performed and reported by assigned personnel for core and important data. Global data sharing and international collaborative research take a back-leap since DSL stringently prohibits sharing Chinese data without prior approval across its borders.

The Privacy Act 1988, of Australia lists 13 Australian Privacy Principles (APPs) that guard the collection, usage and sharing of personal information. Children under 16 are legally banned from social media access and there are strict penalties for data breaches and unlawful sharing of personal information.

Act on the Protection of Personal Information (APPI) in Japan, enforced by the Personal Information Protection Commission, in 2022 requires businesses to ask explicit consent and implement security measures for the safeguard of personal data. The individuals reserve rights for modification and deletion of their data and the Cross-border data transfers are stringently protected.

Lei Geral de Proteção de Dados (LGPD), enforced in Brazil from 2020, by the Autoridade Nacional de Proteção de Dados, outlines ten principles for data safeguard and accountability. The principles apply to all organizations within or outside Brazil that are engaged in processing personal data or offering goods or services to individuals in Brazil. The individuals have been granted rights for correction, deletion, and data portability.

1.4.5 A Final Note on Regulatory Implications

Cross-Cutting Themes.

Across jurisdictions, a clear trend is emerging: erasure obligations are increasingly understood to apply not only to raw records in databases, but to all downstream uses of personal data, including machine learning models, derived representations, and analytics pipelines. Alongside this shift is a growing regulatory and technical demand for provable erasure or unlearning, supported by formal guarantees and measurable evidence, rather than ad hoc or purely procedural database deletions.

Practical Connection.

This urgency is grounded in enforcement reality, as illustrated by landmark decisions such as Google Spain SL v AEPD, which established the Right to Be Forgotten in practice, and substantial GDPR fines against Meta for data protection failures, as well as ongoing CCPA enforcement actions in the United States. These cases signal that regulators expect demonstrable control over how personal data persists and is reused in automated systems.

Future Outlook.

Looking ahead, emerging AI-specific regulations—particularly draft and evolving frameworks in the EU and the US—are expected to further strengthen model unlearning obligations, as AI systems and automated inference are increasingly classified as high-risk uses of personal data, making robust, auditable unlearning mechanisms a core requirement for compliant AI deployment.

1.5 Conclusion

With machine learning gaining momentum across real-world applications, the concerns about how it handles personal data also continue to rise. Sophisticated ML applications specially work well by learning from private or sensitive information, prompting the development of a field called machine unlearning–designed to safeguard users' personal data from abuse. Machine unlearning is a process to forget the influence of specific data points from a model as though they were never seen by the model. It is also essential for maintaining privacy compliance and relevance in evolving ML systems.

It is desirable to adopt modular, privacy-preserving, and scalable frameworks like SAMUL, which offer unlearning on individual data samples, user-level contributions (as in federated learning), or entire tasks without disturbing other capabilities. In its current form, the growth of machine unlearning is faced with challenges, involving trade-offs between efficiency and performance, while guaranteeing that forgotten data leaves no trace. Unlearning evaluation strategies must balance effectiveness with efficiency, incorporating model behaviour comparisons, privacy risk analysis, and performance metrics.

Global data privacy laws like Europe's GDPR, California's CCPA, and regulations across India, China, Australia, Brazil, and Japan underscore the global shift toward data control and accountability. These regulations enforce user rights such as data access, correction, and deletion–making machine unlearning not just a technical innovation, but a compliance necessity in the modern digital era. Machine unlearning is poised to have its greatest impact in high-stakes domains such as healthcare, finance, autonomous vehicles, large language models, and federated learning, where errors in data retention can translate directly into legal, safety, or ethical harm. Researchers, practitioners, and policymakers should collaborate to establish verifiable, benchmarked unlearning pipelines as a new norm, rather than treating unlearning as an ad hoc post hoc fix. Encouragingly, the ecosystem is maturing: open-source unlearning libraries, public leaderboards, and reference datasets—frequently highlighted in recent arXiv surveys and community hubs such as Perplexity Spaces—offer concrete entry points for experimentation, comparison, and the development of shared standards.

Looking Ahead, machine unlearning remains an active and rapidly evolving research area, with open questions around how to evaluate forgetting at scale, how to support unlearning in federated and large language model settings, and how to provide verifiable guarantees without prohibitive retraining costs. As regulatory pressure increases, there is growing potential for standardized benchmarks, certification frameworks, and audit protocols for unlearning. Together, these trends point toward the emergence of "right to be forgotten" engineering as a distinct professional and research discipline at the intersection of machine learning, privacy law, and systems engineering.

Chapter 2
Exact Machine Unlearning Techniques

Abstract This chapter provides a technical analysis of Exact Machine Unlearning (EMU), a field dedicated to the complete and verifiable elimination of data from trained models. We commence by formally defining EMU, emphasizing the distinct theoretical and practical differences from widely utilized approximate unlearning techniques, and elucidating the specific advantages EMU provides regarding certified deletion guarantees. Evaluation of the simplicity, effectiveness, and inherent limitations associated with the basic naive retraining approach is discussed. Subsequently, we introduce the Sharded, Isolated, Slice, and Aggregate (SISA) framework, detailing its architecture and mechanisms designed for promoting efficient, data-partitioned unlearning. Further, the chapter systematically examines state-of-the-art exact algorithms, including DaRE, HedgeCut, and DynFRS, which are explicitly designed for Random Forest models, in addition to the general-purpose framework of ARCANE. Our discussion focuses specifically on partitioning- and tree/ensemble-based EMU approaches and does not attempt an exhaustive taxonomy of all existing methods. This chapter provides a comparative analysis of latency, storage overhead, and mathematical guarantees across these specialized implementations. For each method, we analyze fundamental principles and specific processes to provide a roadmap for executing verifiably exact unlearning.

2.1 What Is Exact Machine Unlearning

Machine learning models, often trained on large amounts of data, are now concerned with security, privacy, and compliance regulations such as GDPR and CCPA, which require specific data points to be removed. The ability to remove specific data points from an already trained machine learning model is referred to as **machine unlearning**. The removal may be complete and may provably erase the influence of specific training sample(s) from an already deployed model, referred to as **exact machine unlearning EMU**, or it may reduce the influence of the data point to be removed, referred to as **approximate unlearning methods**. EMU ensures the resulting unlearned model is mathematically indistinguishable from a model that was trained from scratch solely on the remaining data.

R. Gupta et al., *Machine Unlearning: Concepts, Techniques and Applications*,
Studies in Computational Intelligence 1270,
https://doi.org/10.1007/978-3-032-18447-4_2

Scope of this Chapter This chapter is focused on three primary families of EMU: partitioned training, RF-based structural EMU, and ensemble architectures. Specifically, we detail the SISA framework as the leading approach for partitioned training, which utilizes sharding and slicing to isolate data influence. We also examine structural EMU methods specialized for Random Forests, including DaRE, Hedge-Cut, and DynFRS, which achieve efficiency by localizing updates to specific tree nodes and statistics. Finally, we cover the ARCANE architecture, which represents an ensemble approach using class-wise partitioning and One-Class Classification (OCC) sub-models. While these families cover significant practical implementations, optimization-based EMU for convex Empirical Risk Minimization (ERM) and certain specialized deep-model EMU directions are considered out-of-scope for this discussion, but are referenced throughout for completeness.

2.1.1 Mathematical Definition of Exact Machine Unlearning

Formal Notation: Let $\mathcal{D} = \{(x_i, y_i)\}_{i=1}^{N}$ be the original training dataset, where $x_i \in \mathcal{X}$ represents input features, and $y_i \in \mathcal{Y}$ represents labels. A machine learning model is trained using the following function:

$$\mathcal{M} : \mathcal{D} \rightarrow \Theta$$

where Θ represents the space of model parameters. Given a subset of data $\mathcal{D}_r \subseteq \mathcal{D}$ to be removed, exact machine unlearning requires the model θ obtained after removal to be **identical** to a model trained from scratch on $\mathcal{D} \setminus \mathcal{D}_r$:

$$\mathcal{M}(\mathcal{D} \setminus \mathcal{D}_r) = \mathcal{U}(\mathcal{M}(\mathcal{D}), \mathcal{D}_r)$$

where $\mathcal{U}(\cdot, \mathcal{D}_r)$ is the unlearning operation that removes the influence of $\mathcal{D}_r$. Exact unlearning holds if:

$$\theta^{=} \mathcal{M}(\mathcal{D} \setminus \mathcal{D}_r)$$

which implies that the unlearned model is **indistinguishable** from a model that never saw $\mathcal{D}_r$. This ensures perfect removal of $\mathcal{D}_r$.

2.1.2 Real-World Applications and Scenarios

The implementation of Exact Machine Unlearning (EMU) is essential in high-stakes sectors where the legal burden of proof requires a model to be identical to the retrained model under the same training pipeline. The following examples map the specific technical requirements for exactness across key industries:

1. Financial Institution/Customer: To satisfy the "Right to Erasure" under regulations like GDPR or CCPA, organizations must ensure that sensitive transaction data cannot be reconstructed or detected via Membership Inference Attack (MIA). Approximate methods are insufficient because they only offer empirical similarity, whereas EMU provides a certified mathematical guarantee that the model contains no residual influence from the deleted data.
2. Healthcare Provider/Patient: In medical AI, the total erasure of diagnostic records is required once a patient withdraws consent to ensure future medical decisions remain untainted by that specific data. EMU is necessary to avoid legal liability, as any residual memory in the model parameters could be categorized as unauthorized data retention.
3. Security Systems (Defense/Authentication): Security-critical applications must perform exact unlearning to completely purge the influence of identified poisoned or adversarial samples. Unlike approximate methods, EMU ensures parameter-level erasure, preventing attackers from exploiting the subtle, incremental influences that a single malicious data point can have on a learned representation.

2.1.3 Differences Between Exact and Approximate Unlearning

Machine unlearning techniques are broadly categorized by the rigor of their data removal guarantees. **Exact Unlearning** is mandatory in high-stakes sectors—such as healthcare, finance, and defense—where legal mandates (e.g., GDPR's "Right to Erasure") require a mathematical guarantee that the model is identical to one retrained from scratch. Conversely, **Approximate Unlearning** is acceptable for lower-risk or resource-constrained applications, such as recommendation systems or behavior analytics, where scalability is prioritized over perfect forgetting. While EMU utilizes structural partitioning to isolate data influence, approximate methods typically leverage influence functions or gradient-based unlearning (e.g., gradient ascent) to reduce, but not eliminate, the statistical impact of a target sample (Table 2.1).

2.1.4 Naive Retraining

Naive retraining is the most straightforward approach to achieving exact machine unlearning and serves as the gold-standard baseline for the field. In this process, the original model is discarded entirely, and a new model is trained from scratch using the remaining dataset ($D \setminus D_r$). Figure 2.1 contrasts naive retraining with the localized EMU approaches discussed in later sections. As depicted, naive retraining follows a global path, necessitating a complete re-processing of the remaining data to produce a new model version. In contrast, the EMU path illustrates a targeted unlearning

Table 2.1 Comparison of exact and approximate machine unlearning

Dimension	Exact machine unlearning (EMU)	Approximate unlearning
Primary goal	Produce a model M' mathematically identical to a model retrained from scratch on $D \setminus D_r$	Produce a model M' empirically similar in prediction and loss to a retrained model
Guarantee type	Certified/Provable: Provides a strong theoretical guarantee of zero residual influence	Empirical: Measured by metrics like Membership Inference Attack (MIA) success rates
Typical techniques	Data Partitioning (SISA, ARCANE), Specialized Tree Models (DaRE, HedgeCut, DynFRS)	Gradient Ascent, Influence Functions, Fine-tuning, and Amortized delta-based methods
Applicability	Required when: Strong removal is legally mandated (Healthcare, Finance, Defense)	Acceptable when: Scalability is needed in low-risk scenarios (Content filtering, User analytics)
Verification	Verified through direct parameter-level comparison or formal methods	Verified via statistical tests and membership inference auditing
Use case example	Purging poisoned samples in security-critical authentication systems	Efficiently "rolling back" specific user updates in Federated Learning environments

operation that modifies the state of the existing model to achieve an identical result at the parameter-level without the overhead of a complete rebuild from scratch.

2.1.5 Challenges of Naive Retraining

However, the task of naive retraining has several challenges that make its implementation difficult in practical systems (Fig. 2.2). Such challenges are listed below:

1. **High computational cost**: The most reliable approach, retraining the model from scratch without the target data, is resource-intensive and often impractical for large models or time-sensitive applications.
2. **Model architecture constraints**: Many deep learning models are not designed with modularity in mind, making it difficult to isolate and reverse the effect of specific data points. For modern deep learning models, a full training cycle involves processing the entire dataset over numerous epochs, requiring a massive expenditure of FLOPs (floating-point operations) and significant time for convergence.

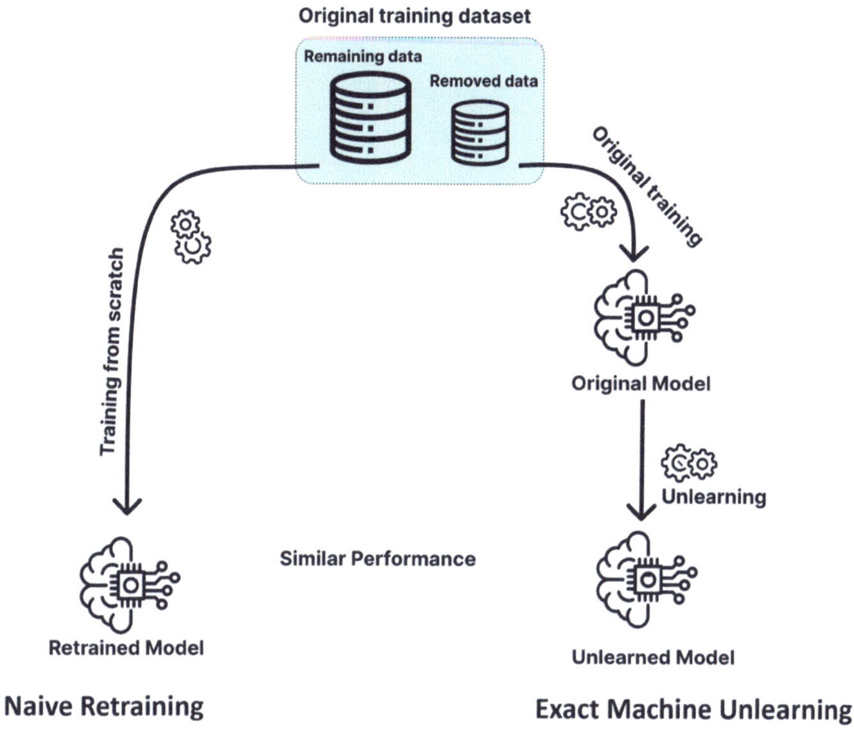

Fig. 2.1 Contrast between Naive retraining and localized EMU approaches

3. **Interdependence of data**: Individual training samples influence the model in complex, incremental ways; removing one point can subtly affect the learned representation of others, complicating accurate removal.
4. **Scalability**: Scalability is a concern in systems where users frequently request data deletion, such as large-scale web services, because each unlearning request can require further training, testing, and deployment cycles.
5. **Verification difficulties**: Verifying that a model has fully forgotten a specific data point remains difficult, as current auditing tools, such as membership inference tests, offer limited assurance regarding actual parameter-level erasure.

These resource-intensive requirements provide the primary motivation for partitioned and model-aware EMU architectures, such as SISA and ARCANE, which aim to achieve the same mathematical results with localized, lower-cost updates. Next sections elaborate on the design dimensions and the techniques of EMU.

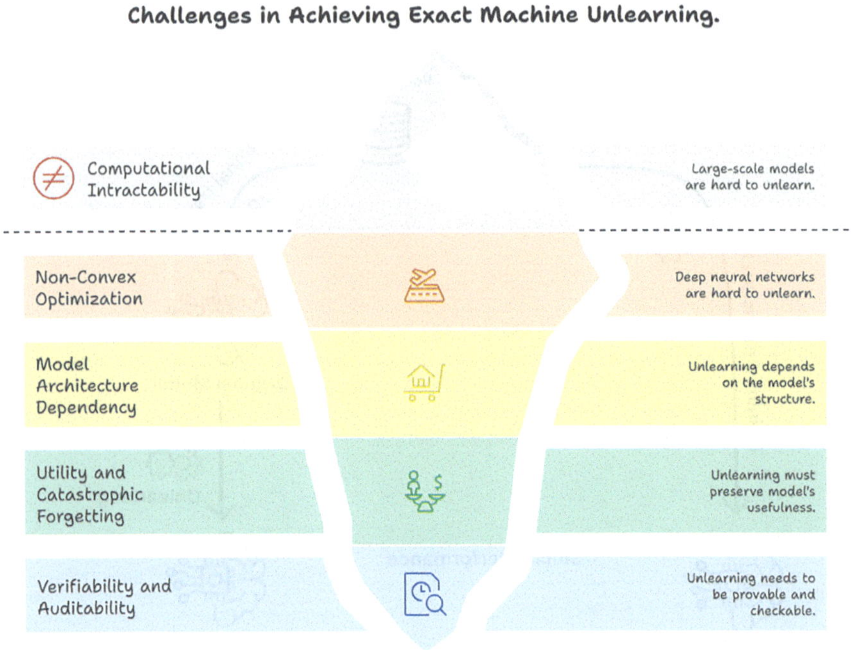

Fig. 2.2 Challenges of exact machine unlearning

2.1.6 *Design Dimensions of EMU Techniques*

We categorize EMU techniques along three primary design dimensions: the unlearning strategy, the operational support, and the model compatibility. This taxonomy enables a structured comparison of how various techniques achieve the goal of mathematical indistinguishability from a retrained model.

2.1.6.1 Taxonomy of Strategies

- **Data Partitioning:** These methods isolate the influence of data points by dividing the training set into disjoint subsets (shards). Unlearning is localized to a specific shard, minimizing the retraining effort. Examples include the general-purpose SISA framework and the class-wise partitioning of ARCANE.
- **Model-Structural Optimization:** These techniques exploit the internal architecture of specific models, most notably tree ensembles, to perform localized updates. By caching statistics at internal nodes or precomputing subtree variants, these methods avoid global retraining. This category includes DaRE, HedgeCut, and DynFRS.

- **Optimization-Based Structure:** This approach leverages the mathematical properties of specific objective functions, such as convex Empirical Risk Minimization (ERM). In these settings, exactness is achieved through algebraic updates to sufficient statistics rather than architectural partitioning or structural caching.

2.1.6.2 Operational and Model Dimensions

Algorithms are further distinguished by their **temporal support**: Online (Real-time) methods focus on microsecond-level latency for immediate compliance (e.g., HedgeCut), while Batch methods optimize throughput for high-volume requests (e.g., DynFRS, SISA with modifications). Furthermore, **Model Compatibility** ranges from Model-Agnostic frameworks like SISA, which can be integrated into most existing deep learning pipelines, to Model-Specific algorithms like the RF-based suite, which are strictly limited to randomized tree ensembles.

2.1.6.3 EMU Design Space Map

Table 2.2 places the core techniques discussed in this chapter within this design space, highlighting their technical trade-offs.

Table 2.2 Taxonomy and design space of EMU techniques

Technique	Category	Model family	Primary support	Key mechanism
SISA	Data-Partitioning	General (DNN/RNN)	Online/Batch	Sharding and Slicing
ARCANE	Data-Partitioning	General (DOCC)	Batch/Online	Class-wise Shards
DaRE	Model-Structural	Random Forests	Online	Statistic Caching
HedgeCut	Model-Structural	ERTs	Real-time	Subtree Variants
DynFRS	Model-Structural	Random Forests	Batch/Online	Lazy Tagging (LZY)
Convex ERM	Optimization	Convex Models	Real-time	Sufficient Stats

2.2 Techniques in Exact Unlearning

We describe several key techniques of EMU in the following sections:

2.2.1 *The SISA Framework*

The Sharded, Isolated, Sliced, and Aggregated (SISA) framework is a foundational architecture for exact machine unlearning designed to mitigate the computational burden of naive retraining through strategic data partitioning and incremental training (Bourtoule et al. 2021). As illustrated in Fig. 2.3, the framework decomposes a global training task into smaller, independent sub-tasks, allowing the system to localize the impact of an unlearning request to a specific subset of model parameters.

2.2.1.1 Core Idea

The core philosophy of SISA is to divide the training dataset D into S disjoint shards and train a separate constituent model M_k for each shard in complete isolation. By further dividing each shard into R slices and saving checkpoints (parameter states)

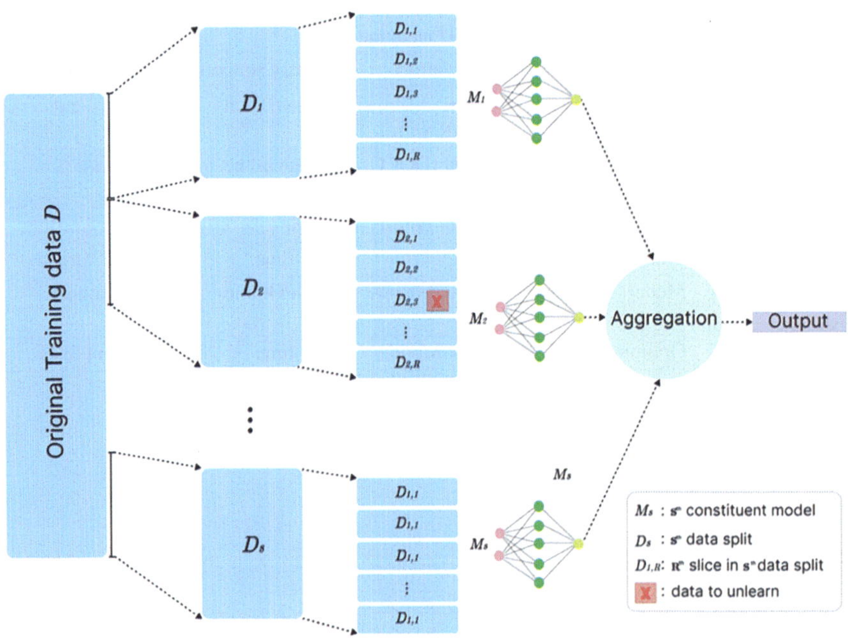

Fig. 2.3 SISA framework

after training each slice, the framework ensures that any unlearning request only necessitates retraining a small fraction of the total model architecture starting from a pre-saved state.

2.2.1.2 Detailed Steps: Sharding, Isolation, Slicing, and Aggregation

The SISA pipeline follows four distinct operational phases:

1. **Sharding:** The original dataset D is partitioned into S disjoint shards $\{D_1, D_2, \ldots, D_S\}$ such that $\bigcap_{k\in[S]} D_k = \emptyset$ and $\bigcup_{k\in[S]} D_k = D$. A constituent model M_k is then trained solely on the data in D_k.
2. **Isolation:** Each model M_k is trained independently without any flow of information between shards. This ensures the influence of a data point $(x_i, y_i) \in D_k$ is strictly limited to model M_k and cannot propagate to models $M_{j\neq k}$.
3. **Slicing:** Each shard D_k is further partitioned into R disjoint slices $(D_{k,1}, D_{k,2}, \ldots, D_{k,R})$. Training proceeds incrementally: M_k first trains on $D_{k,1}$, saves its parameter state, then resumes training on $D_{k,1} \cup D_{k,2}$, and so on. Upon a request to unlearn data point $d_u \in D_{k,r}$, only the affected model M_k is retrained, starting specifically from the parameter state saved immediately before slice $D_{k,r}$ was introduced.
4. **Aggregation:** For inference, the framework collects predictions from all S constituent models. A final prediction is produced via an aggregation strategy, most commonly a majority vote on the predicted labels.

2.2.1.3 Assumptions and Modifications

The standard SISA framework relies on specific assumptions regarding the nature of data and requests:

- **Uniform Distribution Assumption:** In the original SISA proposal, it is assumed that data points to be unlearned are equally likely to be found across all shards, and shards are created via uniform random partitioning.
- **Sequential Processing:** Requests are assumed to be handled individually as they arrive, rather than being collected in batches.

Modifications for Non-uniform Scenarios: If prior knowledge of the unlearning distribution is available (e.g., certain user roles or data types are more likely to be deleted), the framework can be optimized through various strategies (Koch and Soll 2023) which are listed below:

- **Distribution-Aware Sharding:** This modification is utilized when the unlearning likelihood of specific data points can be estimated in advance (e.g., for high-unlearning user segments). By placing points with high erasure probabilities into designated shards, the system ensures that frequent retraining is localized to a small number of sub-models. While this optimizes the expected retraining cost, it introduces a trade-off regarding model generalization, as these shards may become highly biased or fail to capture the full data distribution, leading to weaker constituent learners.

- **Distribution-Aware Slicing:** Within a specific shard, retraining efficiency is maximized by placing high-risk data points into later slices ($D_{k,r}$ where r is large). Because the unlearning process only requires retraining from the checkpoint immediately preceding the affected slice, this strategy minimizes the number of subsequent slices that must be re-processed, significantly reducing the total computational time per request.
- **Batch Unlearning:** Rather than addressing each request in sequence, the system may buffer multiple unlearning requests and process them as a single batch. This approach significantly increases system throughput and reduces the cumulative retraining overhead, although it comes at the expense of higher per-request latency. Crucially, batching maintains the exactness of the unlearning guarantee, as the resulting model remains mathematically identical to one retrained on the reduced dataset.

2.2.1.4 Empirical Behavior and Trade-Offs

The effectiveness of SISA was evaluated by (Bourtoule et al. 2021) across various datasets and model complexities, using full retraining from scratch as the baseline.

- **Simple Learning Tasks (Purchase and SVHN):** Using a configuration of 20 shards and 50 slices, SISA achieved a speed-up of 4.63x for the Purchase dataset and 2.45x for SVHN. In these scenarios, the accuracy degradation was nominal, remaining under 2% points.
- **Complex Tasks (ImageNet):** For deeper networks like those used for ImageNet, the speed-up was lower at 1.36x. While initial experiments showed a significant performance drop (19.45% point loss in top-5 accuracy for 39 requests), this was successfully mitigated to a 2% point loss by coupling SISA with transfer learning.
- **The Storage Trade-off:** While slicing consistently provides unlearning speed-ups, it introduces a substantial storage overhead because the system must maintain multiple intermediate parameter states (checkpoints) for every shard.

2.2.2 Random Forest Based EMU

A random forest, an ensemble learning method, is a collection of individual decision trees (sub-models) on the subset of the dataset (shard). The influence of the data point is limited to the decision tree in which it exists. Retraining the random forest will be limited to retraining of the relevant decision tree only. The prediction of the model is based on an aggregate of the predictions of individual decision trees.

The structure of a random forest (RF) is similar to the SISA structure, with the difference that no slicing is done in the case of RFs. Further, each tree in a random forest can be viewed as a constituent model; however, unlike SISA, RF shards overlap in feature and sample space. Several SISA-based approaches like DaRE (Brophy and Lowd 2021), HedgeCut (Schelter et al. 2021), DynFrs (Wang et al. 2025) are based on random forest framework, which exploits local statistics and structure in trees to localize updates to affected subtrees.

2.2.2.1 DaRE: Data Removal Enabled Forests

A DaRE Forest is an ensemble of independently trained DaRE Trees that utilizes a hybrid, two-level architectural approach to balance stability and predictive power (Brophy and Lowd 2021).

Training Scheme:

- **Upper Levels (Random Nodes):** Splits are chosen by selecting both the feature and the threshold uniformly at random. These nodes are designed to have limited data-dependence, providing structural stability near the root and avoiding costly cascading updates during the unlearning process.
- **Lower Levels (Greedy Nodes):** These nodes utilize standard greedy splits optimized based on measures like Gini index or entropy. They function as traditional decision tree nodes but are augmented to store local training data and cached statistics at each node and leaf.

The Unlearning Process To remove an instance (z) from a DaRE Forest, execute the following steps for every affected tree:

- **Traverse and Update Statistics:** Starting from the root, follow the path to the leaf where the deleted instance z is located. At every node along this path, update the cached statistics (e.g., instance counts, split scores) by subtracting the contribution of z.
- **Update the Leaf:** Upon reaching the terminal node, remove the instance z from the stored training data pointers and recompute the leaf's prediction value.
- **Evaluate Subtree Rebuilding:** At each internal node on the path, use the updated statistics to re-evaluate the optimal split criterion. If the deletion causes the optimal split to change, trigger a full re-train of the specific subtree originating from that node using the remaining data.

Unlearning Structure Summary:

- **Inputs:** The specific data instance z to be forgotten.
- **Precomputed Metadata:** Cached statistics, including instance counts, split criteria data, and instance pointers stored at every node.

- **Update Path:** The direct path from the tree root to the specific leaf containing instance z.
- **Stopping Condition:** The process terminates if the optimal split criterion at an internal node remains unchanged after the statistical update, or when a leaf is reached.

Experimental Results Experimental evaluations confirm that DaRE Forests provide a massive efficiency gain, with data deletion typically occurring 2–4 orders of magnitude faster than the simple approach of retraining the entire model from scratch. While DaRE maintains accuracy nearly identical to a standard Random Forest, it achieves sub-second latency for most requests; for instance, it has been reported to handle updates with a latency of approximately 180 ms, compared to the much longer durations required for full retraining.

Strengths and Weaknesses The primary advantages of DaRE are its guaranteed exactness and high computational efficiency due to localized updates. However, these benefits come with a significant memory cost, as the algorithm requires 6x–26x more storage than a standard Random Forest to maintain the necessary cached statistics and instance pointers. Additionally, the framework is model-specific and requires careful hyperparameter tuning of the random node depth to balance accuracy and unlearning speed.

2.2.2.2 HedgeCut: Maintaining Randomised Trees for Low-Latency Machine Unlearning

HedgeCut is an in-place, low-latency unlearning method designed for production environments where deletion requests must be handled in milliseconds or even microseconds (Schelter et al. 2021). It is built upon an ensemble of Extremely Randomized Trees (ERT), where the split attributes are chosen randomly at each node to decouple split decisions from the specific training data.

Training Scheme:
HedgeCut manages the structural stability of the trees by distinguishing between two types of split decisions during the training phase:

- **Robust Splits:** These are split decisions quantified (e.g., via Gini gain analysis) as statistically unlikely to change even after the removal of a small subset of records. These splits are treated as fixed, allowing the algorithm to skip expensive structural checks during unlearning.
- **Non-Robust Splits:** These splits are identified as sensitive to data removal. For these nodes, HedgeCut precomputes alternate subtree variants during initial training. This ensures that if a deletion affects the validity of the primary split, an "unlearned" version is already available for immediate use.

Data Structures and Trade-offs To achieve near-instantaneous unlearning, HedgeCut maintains a complex set of internal data structures, leading to a significant time versus memory trade-off:

- **Maintained Structures:** Every node and leaf stores cached statistics (e.g., class counts) and instance pointers. Additionally, non-robust nodes store variant pointers to precomputed alternate subtrees.
- **Cost Trade-off:** The algorithm provides unprecedented low latency (approximately 100 μs), supporting massive request throughput. However, this speed comes at the cost of a high memory footprint, as storing statistics and subtree variants for even a small fraction of nodes (typically $< 2\%$) increases system requirements.

The Unlearning Process When a request is received to forget data point z, HedgeCut performs the following sequence:

1. **Identification:** Identify all trees in the ensemble that were trained using instance z.
2. **Path Traversal and Subtraction:** For each affected tree, traverse the path from the root to the leaf containing z. At every internal node, update the cached statistics by subtracting the contribution of z.
3. **Subtree Switching (for Non-robust Nodes):** At any node along the path identified as non-robust, the algorithm evaluates if the deletion of z triggers a need to change the split. If so, it switches to the pre-calculated alternate subtree variant rather than rebuilding the tree.
4. **Fixed Split Optimization (for Robust Nodes):** For robust nodes, the algorithm utilizes its Fixed Splits Guarantee to bypass structural re-evaluation, performing only the statistical subtraction.
5. **Leaf Recomputation:** Upon reaching the leaf, recompute the prediction label based on the updated, z-free statistics.
6. **In-Place Update:** The tree is updated instantly in production, ensuring the model remains mathematically consistent with one trained from scratch.

Experimental Results Experimental evaluations on several sensitive, real-world datasets confirmed the success of HedgeCut in achieving its low-latency objective. The algorithm was demonstrated to perform instance unlearning in approximately 100 microseconds (μs), supporting a massive throughput of prediction and deletion requests. Furthermore, the predictive accuracy and initial training time of the HedgeCut model were shown to be highly competitive with, or similar to, standard and well-established models like Extremely Randomized Trees. This validates HedgeCut as a highly practical solution for production environments that require the mandatory compliance of the Right to be Forgotten.

Strengths and Weaknesses The primary advantage of HedgeCut is its millisecond-level latency, which is ideal for high-throughput services with mandatory real-time compliance requirements. The reliance on precomputed variants and fixed splits simplifies unlearning to fast statistical subtraction. However, its scope is limited to

Extremely Randomized Trees, and it is not directly applicable to models like deep neural networks. Furthermore, the increased storage requirement for variants and pointers remains a bottleneck for memory-constrained systems.

2.2.2.3 DynFRS: An Efficient Framework for Machine Unlearning in Random Forests

DYNFRS (Dynamic Forest Re-Sampling) is a specialized framework designed to provide high-speed, exact machine unlearning and continual learning (data addition) for Random Forest models (Wang et al. 2025). It is specifically optimized for high-scale environments requiring real-time compliance with privacy regulations.

Training Scheme:
The framework is built upon two synergistic strategies that allow it to bypass the heavy computational costs associated with immediate tree reconstruction:

- **Occurrence Subsampling (OCC(q)):** This method ensures that any single training sample is used to train only a maximum proportion (q) of the trees in the forest. This limits the blast radius of a deletion request to a small, predictable subset of trees.
- **Lazy Tag Strategy (LZY):** Instead of immediately recalculating node statistics when data is removed, DYNFRS utilizes a pay-on-demand approach. Affected nodes are simply marked with a "lazy tag," deferring expensive physical reconstruction until the node is actually needed for a prediction.

The Unlearning Process DYNFRS handles unlearning requests through a decoupled process that separates the modification request from the actual model update:

1. **Scope Identification (Modification Phase):** Using the OCC(q) principle, the system immediately identifies the specific subset of trees containing instance z.
2. **Lazy Tagging:** The internal nodes of these trees are marked with a **Lazy Tag (LZY)**, indicating that their current statistics are "stale" due to the removal of z.
3. **Postponed Updates:** The actual data removal and statistical re-calculation are postponed. This ensures the modification request itself completes with microsecond-level latency.
4. **On-Demand Reconstruction (Query Phase):** The actual reconstruction of a lazy-tagged node occurs only when that node is encountered during a prediction query.
5. **Tag Clearance:** Once the node is updated on-the-fly, the lazy tag is cleared, and the node is considered "clean" for future queries.

Experimental Results DYNFRS demonstrates remarkable efficiency in handling modification requests. While traditional methods like DaRE have reported modification latencies of approximately 180 ms, DYNFRS achieves an average modification latency of less than 0.15 ms. This represents an improvement of several orders of magnitude, making it uniquely suited for high-speed, real-time data streams.

Strengths and Weaknesses The primary strength of DYNFRS is its extremely low modification latency and its ability to handle both additions and deletions seamlessly. By shifting the computational burden from the deletion time to the prediction time, it allows systems to remain responsive even under heavy unlearning loads.
However, this pay-on-demand model introduces a query-time trade-off. As the number of unprocessed unlearning requests (lazy tags) increases, query latency may rise because the model must perform reconstructions during active inference. Additionally, the framework requires a non-standard modification of the model structure to support the tagging and on-demand reconstruction logic.

2.2.2.4 Comparison of Random Forest-Based Methods

Each of the random forest based method for exact unlearning has its merits and demerits. Table 2.3 summarizes these points.

2.2.3 ARCANE: Architecture for Exact Machine Unlearning

ARCANE, based on ensemble learning, is a novel architecture designed to address the inefficiency of naive re-training by dividing the task into multiple One-Class Classification (OCC) tasks (Yan et al. 2022). Data preprocessing steps like redundancy removal using representative data selection and the reuse of previous calculation results through training state saving reduce the retraining overhead and speed up the entire process. Further, sorting as a preprocessing step speeds up the retraining if prior knowledge of the probability distribution of the unlearning data is known. Final prediction is made by aggregating the outputs of all OCC sub-models, such as via majority vote or max confidence. The approach is considered suitable for cases where a large number of unlearning requests are expected, specifically more than 10% of the training data. The use of the one-class concept makes it quite effective for unbalanced datasets as well.

ARCANE Architecture The core of the architecture relies on the One-class classifier (OCC). The dataset is split on the basis of class labels into n disjoint shards (D_k), where each shard contains data for a single class k, ranging from 1 to n. This strict class-wise partitioning ensures that each shard has data corresponding to only one class, a method different from SISA where shards are created randomly. ARCANE will always produce exactly n number of shards, with each shard trained separately as an independent sub-model.
To implement these sub-models, ARCANE utilizes Deep One-Class Classification (DOCC) techniques, such as Deep Support Vector Data Description (SVDD), to accurately represent the retained data within each small shard. By replacing resource-intensive multi-class classification retraining with the much faster DOCC task, the system achieves a massive efficiency boost for exact unlearning.

Table 2.3 Comparison of random forest-based EMU methods

Feature	DaRE (data removal-enabled)	HedgeCut	DYNFRS (dynamic random forest)
Core idea	Retrains subtrees only as needed by caching statistics at nodes and training data at leaves; utilizes random nodes at top levels to minimize structural dependency on specific samples	Uses split robustness to identify nodes unlikely to change; maintains pre-calculated subtree variants for non-robust splits to allow instant replacement upon data removal	Combines $OCC(q)$ subsampling to limit sample impact to specific trees with a lazy tag strategy (LZY) that delays structural updates until a query request arrives
Time cost for deletion	Sub-second latency; typically 2–4 orders of magnitude faster than retraining from scratch	Extremely low latency (around 100 microseconds); designed for near-instant responses in serving systems	Very low latency (avg. 0.12 ms); significantly faster than DaRE for sequential unlearning and specializes in high-speed batch unlearning
Storage overhead	High overhead (6x–26x relative to standard RF) due to caching attribute-threshold statistics and instance pointers	Requires additional memory to store subtree variants for non-robust splits, though such nodes are typically a small fraction ($<2\%$) of the tree	Moderate overhead; consumes less memory than DaRE by using subsampling and storing split statistics rather than full instance lists for every node
Model scope	Random Forests (DaRE Forest variant).	Extremely Randomized Trees (ERTs)	Random Forests (adaptable to any variant, including ERTs)
Best suited scenarios	Scenarios requiring exact unlearning for Random Forests where many sequential deletion requests are expected	Deployed serving models requiring millisecond-level unlearning latency for a small fraction (e.g., 0.1%) of the dataset	High-scale environments requiring efficient batch unlearning or real-time mixed data streams (addition, removal, and query)

Data Pre-processing Reducing the data for training speeds up the entire process. These preprocessing steps are tightly connected to reducing computational complexity while maintaining the exactness of the final model.

- **Representative Data Selection:** Redundant or irrelevant data can slow down the training process. ARCANE utilizes information theory to identify and remove such data by calculating joint entropy on the latent representations of each sample to assess its information content. While samples holding more information are retained to expedite retraining without sacrificing accuracy, there is a potential risk of discarding important boundary points if the selection is too aggressive.

- **Training State Saving:** This feature improves unlearning efficiency by establishing checkpoints throughout the training process of the isolated DOCCs. Rather than training a single final model, ARCANE retains intermediate model versions. When an unlearning request is issued, only the relevant OCC is retrained, beginning from the most recently saved state instead of starting from the beginning, which minimizes computation time.
- **Sorting by Erasure Probability:** This advanced optimization involves pre-analyzing the dataset to estimate the likelihood that each data point will be subject to a future unlearning request. In practice, this probability might be estimated using user roles or consent logs. By strategically positioning data with a higher probability of erasure later in the training sequence or within easily isolated partitions, the system ensures the model only needs to revert to a very recent checkpoint upon deletion. While effective, this sorting must be handled carefully due to potential privacy implications regarding the storage of deletion likelihoods.

Unlearning Process The unlearning process guarantees exact unlearning by leveraging its modular, class-based architecture. The correctness of the EMU relies fundamentally on: (1) strict class-wise partitioning and (2) the fact that deleted points affect only their specific class-shard.

1. **Request Localization ($D_f \rightarrow M_k$):** Upon receiving a request to unlearn a forget set D_f, ARCANE identifies the specific class k associated with D_f. Due to the class-based partitioning and isolated training, the influence of D_f is entirely contained within the single OCC M_k.
2. **Targeted Retraining:** Only the affected sub-model M_k is retrained on the reduced dataset $D_k \setminus D_f$, while all other sub-models remain unchanged.
3. **Efficiency Boost:** Retraining is accelerated by the Training State Saving mechanism, allowing M_k to resume from the latest checkpoint prior to the inclusion of the deleted data.

Experimental Results Experimental evaluation across image datasets (CIFAR-10/100) and privacy-sensitive purchase data demonstrated that ARCANE achieves 100% exact unlearning while maintaining test accuracy comparable to gold-standard models. ARCANE exhibited a superior speedup compared to naive re-training and improved efficiency over the SISA baseline, particularly in scenarios with high volumes of deletion requests (e.g., 50% of the data), making it a highly scalable solution.

Strengths and Weaknesses ARCANE's primary strength is its massive efficiency boost for exact unlearning. This approach is highly suitable for production systems that need to handle frequent unlearning requests without long periods of model downtime. However, the overall performance of the aggregated model is generally not as high as that of a single, monolithic model trained on the complete dataset. The success of the method is also dependent on the effectiveness of the chosen DOCC technique in accurately representing the data within each shard.

2.3 Comparison of Experimental Results of EMU Techniques

Table 2.4 compares the experimental results of various EMU techniques discussed above.

Summarizing the above table, Random Forest-based suite (HedgeCut and DynFRS) offers the lowest latency, achieving microsecond-to-millisecond performance by localizing updates to specific tree nodes. SISA and ARCANE provide broader applicability for deep learning models, with ARCANE demonstrating particular efficiency in high-volume deletion scenarios due to its class-wise partitioning. Most methods maintain performance within 2% points of the gold-standard retraining baseline, although some degradation is noted in complex tasks or for minority classes. Based on these findings, the next section suggests choosing an EMU Technique in various scenarios.

2.3.1 Choosing an EMU Technique in Practice

Selecting an EMU algorithm requires a careful balancing of the model architecture, the volume of unlearning requests, and the permissible latency for model updates. While framework-level solutions like SISA provide broad applicability, model-specific optimizations like DynFRS or HedgeCut are often necessary to meet the stringent latency requirements of production serving environments. The following Table 2.5 serves as a practitioner's guide for navigating these trade-offs.

2.3.2 Verification and Auditing of EMU Techniques

While empirical auditing tools such as Membership Inference Attacks (MIA) and statistical divergence tests are frequently used to evaluate approximate unlearning, they face significant limitations when applied to EMU. Specifically, these tools offer limited assurance regarding actual parameter-level erasure, as they primarily measure output behavior rather than internal weight states.

Consequently, for the partitioned and structural methods detailed in this chapter, including SISA, ARCANE, and the RF-based algorithms, the guarantee of exactness is predominantly analytical or proof-based. This "certified deletion" ensures that the unlearning process is mathematically consistent with the original training algorithm by construction, providing a level of verifiable compliance that purely empirical or heuristic approaches cannot offer.

Table 2.4 Experimental results comparison of EMU techniques

Method	Datasets	Baseline	Metrics	Quantitative results
SISA	Purchase, SVHN, ImageNet	Naive retraining from scratch	Speed-up, Test accuracy	4.63x (Purchase) and 2.45x (SVHN) speed-up with <2% accuracy loss. ImageNet speed-up 1.36x with accuracy loss mitigated to 2% via transfer learning
DaRE	Various general datasets	Naive forest retraining	Deletion latency, Accuracy	2–4 orders of magnitude faster than retraining from scratch; reported latency of ~180 ms. Accuracy is comparable to standard Random Forests
HedgeCut	Sensitive real-world datasets	Standard Extremely Randomized Trees (ERT)	Unlearning latency, Accuracy	Near-instant unlearning with 100 μs latency. Predictive accuracy remains highly competitive with standard ERTs
DynFRS	Higgs (massive scale)	DaRE	Modification (add/delete) latency	Average modification latency of 0.12 ms (significantly faster than DaRE's 180 ms). Highly scalable for massive datasets
ARCANE	CIFAR-10, CIFAR-100, Purchase data	Naive retraining, SISA baseline	Speed-up, Test accuracy	Superior speed-up over both SISA and naive retraining, particularly at high deletion volumes (e.g., 50% of data). Accuracy is comparable to gold-standard models

Table 2.5 Practitioner's guide: selecting an exact machine unlearning technique

Scenario	Primary constraints	Recommended technique
Tabular fraud detection	High accuracy; Random Forest model; sequential deletes	**DaRE:** Optimized for sequential updates in standard RF structures
Real-time serving systems	Near-instant updates (μs); production-ready ERT models	**HedgeCut:** Provides microsecond latency (∼100 μs) for serving environments
High-scale real-time streams	Mixed add/delete requests; massive datasets (e.g., Higgs)	**DynFRS:** Optimized for real-time mixed streams and batch unlearning
Deep learning (images/text)	State-of-the-art models (CNNs, Transformers); moderate volume	**SISA:** Model-agnostic; integrates into existing deep learning pipelines
Highly imbalanced datasets	High volume of deletions (>10%); class-specific requests	**ARCANE:** Uses class-wise OCC shards to handle imbalanced data effectively
Low-risk user analytics	Extreme scale; tolerance for minor residual influence	**Consider Approximate MU:** Scalable for systems like content filtering or recommendations

2.4 Conclusion and Future Directions

Exact Machine Unlearning (EMU) provides the necessary mathematical framework to honor the "Right to Erasure," ensuring that a model's state is indistinguishable from one that never encountered the forgotten data. While naive retraining remains the theoretical gold standard, the specialized methods examined in this chapter, SISA, the Random Forest-based suite (DaRE, HedgeCut, DynFRS), and ARCANE, offer essential pathways toward practical efficiency.

The choice of an EMU strategy should be driven by the specific data modality and operational requirements. SISA is most effective for general deep learning architectures (e.g., RNNs or Transformers) where modular retraining through sharding and slicing is feasible. For tabular data requiring near-instantaneous responses, Random Forest-based methods are superior, with HedgeCut optimized for millisecond-level real-time deletes and DynFRS for high-speed batch operations. Finally, ARCANE is uniquely suited for image classification and regulated domains where high-volume unlearning requests or significantly unbalanced datasets are prevalent.

Despite these advancements, several critical challenges remain for the next generation of EMU research. Scaling these techniques to large-scale deep models remains a primary bottleneck due to the high computational cost and lack of inherent modularity in deep architectures. Furthermore, the field requires more robust verification tools capable of auditing actual parameter-level erasure to move beyond the limited assurance offered by current membership inference tests. Finally, future work must address the fairness of unlearning, specifically how localized retraining can disproportionately degrade accuracy for minority classes, and explore the complex interaction between unlearning and continual learning as models evolve over time.

Chapter 3
Approximate Machine Unlearning Techniques

Abstract This chapter presents a comprehensive exploration of approximate unlearning techniques, addressing the urgent need for efficient data removal from trained machine learning models in response to privacy regulations such as the GDPR's "right to be forgotten." Unlike exact unlearning, which retrains models from scratch at substantial computational cost, approximate unlearning seeks to rapidly eliminate the influence of forgotten data with acceptable fidelity, balancing unlearning efficacy, model utility preservation, and computational efficiency. The chapter systematically categorizes and analyzes state-of-the-art algorithms, including parameter-modification (e.g., Amnesiac Learning, Eraser, DeltaGrad), meta-learning approaches (LTU, SFTC), knowledge distillation, sparsity-aware, and specialized methods for federated, zero-shot, and post-deployment scenarios. Deployment considerations and experimental findings demonstrate that no single method dominates across all metrics; instead, algorithm choice depends on model architecture, dataset properties, operational constraints, and regulatory context. Finally, the chapter identifies persistent challenges—such as robust evaluation metrics, LLM unlearning, and formally certified approaches—and highlights the pivotal role of approximate unlearning in responsible AI as both privacy expectations and model deployment scale rapidly.

Keywords Approximate unlearning · Privacy compliance · Model utility preservation · Data removal efficiency

R. Gupta et al., *Machine Unlearning: Concepts, Techniques and Applications*, Studies in Computational Intelligence 1270,
https://doi.org/10.1007/978-3-032-18447-4_3

3.1 Introduction and Conceptual Foundation

3.1.1 The Machine Unlearning Problem and Regulatory Context

Machine unlearning represents one of the most significant challenges in contemporary machine learning: efficiently removing the influence of specific training data from trained models without complete retraining. This problem emerged from regulatory necessity. The European Union's General Data Protection Regulation (GDPR) and similar privacy laws worldwide have introduced the "right to be forgotten," legally mandating that individuals can request deletion of their personal data from all systems that hold it, including trained machine learning models.

Given an original model θ_o trained on dataset D containing N samples, the unlearning problem seeks to construct an unlearned model θ_u that behaves as if trained exclusively on the remaining dataset $D_r = D \setminus D_f$, where D_f represents the data to be forgotten. The fundamental challenge lies in achieving this objective without complete model retraining, which would incur prohibitive computational costs—especially for modern deep neural networks with billions of parameters.

3.1.2 Exact Versus Approximate Unlearning: The Fundamental Trade-Off

Exact unlearning achieves the theoretical ideal by completely retraining the model from scratch on D_r. The retrained model $\theta_r = \underset{\theta}{\text{argmin}} L(D_r, \theta)$ serves as the gold-standard benchmark representing perfect unlearning. However, this approach demands computational resources proportional to the original training cost, making it impractical for frequent deletion requests.

Approximate unlearning relaxes the exactness requirement by accepting models whose outputs diverge slightly from the exactly retrained model while remaining computationally tractable. The approximation gap quantifies this performance difference between approximately unlearned model θ_u and the exact retrained model θ_r. Empirical results from recent research demonstrate that approximate unlearning methods can achieve 10-100X speedups over complete retraining while maintaining acceptable fidelity to the gold-standard model.

This trade-off creates a three-dimensional optimization landscape:

Unlearning Efficacy: Measured by accuracy degradation on forgotten data D_f and membership inference attack (MIA) success rates. The unlearning accuracy (UA) represents the model's error rate on forgotten samples, with UA close to 100% indicating complete forgetting.

Model Utility Preservation: Quantified through remaining accuracy (RA) on retained data D_r and test accuracy (TA) on the original dataset's test split. High utility preservation ensures the model remains functional for its intended purpose.

Computational Efficiency: Measured in runtime and memory requirements. Approximate methods must demonstrate substantial speedups over retraining to justify their deployment in production systems.

3.1.3 Real-World Deployment Scenarios

GDPR Compliance and Data Deletion: Organizations processing EU citizen data face legal obligations to delete training data upon user request. For platforms with millions of users, complete retraining for each deletion would render the system economically infeasible. Approximate unlearning enables timely deletion while maintaining model performance.

Privacy Breach Remediation: Organizations frequently discover that training data contained improperly collected or sensitive information (medical records, children's data, personally identifiable information). Removing such data through approximate unlearning allows rapid compliance without service interruption.

Intellectual Property and Copyright Removal: Rights holders increasingly request removal of their copyrighted content from large language model training datasets. Approximate unlearning techniques enable efficient removal of specific classes or concepts from trained models without retraining.

Content Safety and Alignment: Platforms must rapidly remove harmful knowledge (instructions for creating weapons, illegal content, malicious information) from production models. Exact unlearning's lengthy retraining timeline makes it unsuitable for urgent safety interventions.

Model Maintenance and Debugging: Detecting corrupted, mislabeled, or poisoned training data requires rapid removal and remediation. Approximate unlearning provides the necessary speed for iterative model debugging and maintenance.

3.2 Approximate Machine Unlearning Algorithms

3.2.1 *Parameter-Modification Methods: Direct Weight Adjustment*

Parameter-modification approaches directly adjust model weights to reduce forgotten data influence. These methods typically optimize composite loss functions balancing unlearning objectives with utility preservation.

3.2.1.1 Amnesiac Learning: Gradient-Based Reverse Updates

Amnesiac unlearning leverages a crucial insight: during original training, model parameters evolve through accumulated gradient updates. The key innovation involves maintaining records of which training examples appeared in which batches and their corresponding parameter updates.

Mathematical Foundation: During standard training with stochastic gradient descent, model parameters evolve as:

$$\theta_t = \theta_0 + \sum_{i=1}^{t} \Delta\theta_i$$

where $\Delta\theta_i$ represents the parameter update applied during batch i.

Amnesiac unlearning subtracts parameter updates from batches containing forgotten data:

$$\theta_u = \theta_o - \sum_{i \in S_f} \Delta\theta_i$$

where S_f denotes all batches containing at least one sample from the forget set.

Advantages: Extremely efficient for removing small numbers of data points with minimal parameter perturbations. Achieves unlearning accuracy near 100% when removing single samples.

Limitations: Requires storing all batch-level gradient updates during training (memory overhead proportional to total training iterations multiplied by model size). Effectiveness diminishes with larger forget sets due to accumulated approximation errors. Requires careful hyperparameter tuning to prevent catastrophic parameter drift.

Experimental Insights: On CIFAR-10 with ResNet-18 removing 10% of data, Amnesiac achieves 7.58% unlearning accuracy gap versus retraining while maintaining 98.87% remaining accuracy. Performance degrades to 14.99% gap with 50% data removal.

3.2.1.2 Eraser: Fisher Information-Based Influence Removal

Eraser extends influence-function theory from classical statistics to deep neural networks. The influence function approximates how a data point affects final model parameters, enabling targeted removal of specific samples' influence.

Theoretical Framework: The approximation $\theta_u \approx \theta_o - H^{-1}\nabla L_f$ relates parameter changes to the Hessian matrix of the training loss and gradient with respect to forgotten data.

Practical Implementation: Direct Hessian computation proves computationally prohibitive for large models. Eraser uses the Fisher Information Matrix (FIM) as an approximation, replacing H with $\mathbb{E}_{(x,y)\sim D}[\nabla_\theta L(x, y)\nabla_\theta L(x, y)^T]$.

Trade-offs: More theoretically grounded than simple fine-tuning, with approximation bounds connecting unlearning error to model properties. However, Fisher approximation requires second-order information computation, which remains expensive for billion-parameter models (Fig. 3.1).

3.2.1.3 DeltaGrad: Cached Gradients and Limited-Memory Approximation

DeltaGrad reduces computational burden through efficient approximation of Hessian-vector products using limited-memory BFGS (L-BFGS) and cached gradient information.

Optimization Strategy: Rather than computing the full Hessian at unlearning time, DeltaGrad maintains gradient information during training:

$$\Delta\theta = -H_t^{-1}\nabla L_f$$

Using L-BFGS approximations with cached gradients dramatically reduces computational requirements.

Key Advantages: Enables both data addition and removal operations. Works effectively with limited access to original training data. Demonstrates practical speedups for federated learning scenarios where data remains distributed.

Implementation Requirements: Maintaining cached gradients throughout training; appropriate tuning of L-BFGS parameters to balance approximation quality with computational cost.

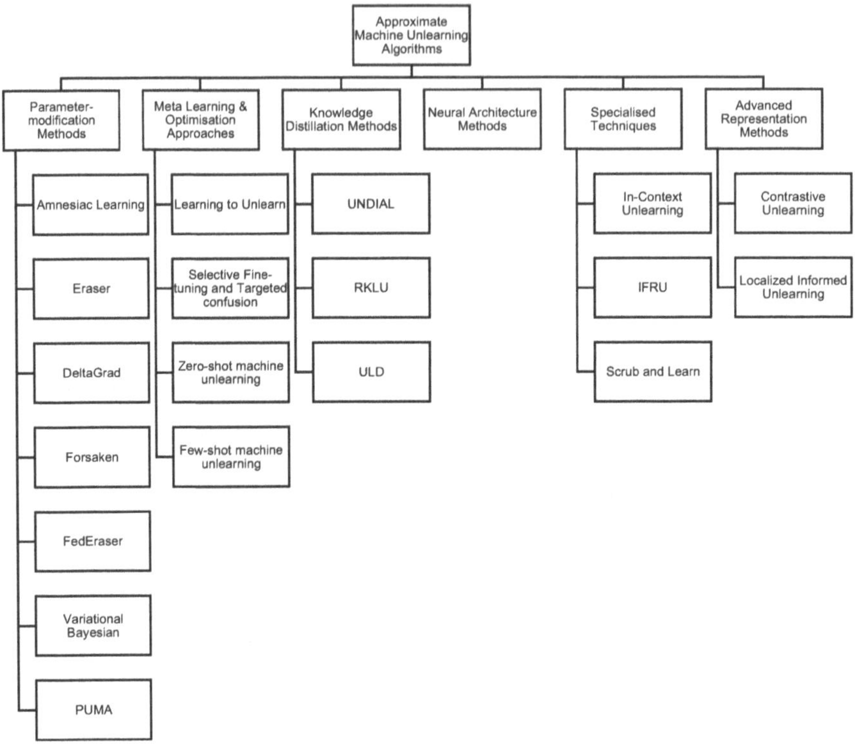

Fig. 3.1 High level categorization of different approximate machine unlearning algorithms

3.2.1.4 Forsaken: Neuron Masking and Bounded Updates

Forsaken introduces neuron-level localization by identifying neurons most relevant to forgotten data and updating only those neurons with bounded gradient steps.

Process Architecture:

1. **Neuron Identification**: Use saliency analysis (gradient-based importance scores) to identify neurons with high activation magnitude on forgotten data
2. **Selective Updates**: Compute gradient updates specifically for identified neurons
3. **Bounded Modification**: Apply updates with magnitude constraints to prevent catastrophic model collapse
4. **Fine-tuning**: Adjust remaining parameters to restore model performance

Benefits: Focuses changes on task-relevant parameters, reducing side effects on retained knowledge. Prevents catastrophic forgetting of general model capabilities. Increases interpretability compared to global parameter modifications.

3.2.1.5 FedEraser: Federated Learning Calibration

FedEraser addresses the complexity of unlearning in federated learning, where training data remains distributed across multiple clients without central access.

Federated Architecture: Maintains calibration factors for each client's historical updates:

$$\theta_u = \theta_o - \sum_{i \in \text{clients with forgotten data}} \gamma_i \Delta\theta_i^{(client)}$$

where γ_i represents learned calibration weights accounting for each client's update influence.

Federated Innovations: Handles sequential updates from multiple clients; prevents data leakage in collaborative learning; enables efficient removal without retraining individual clients. Experimental results show computational savings of 2 × compared to retraining while maintaining 99.45% remaining accuracy on CIFAR-10.

3.2.1.6 Variational Bayesian Unlearning: Posterior Distribution Adjustment

For Bayesian models, unlearning reformulates as minimizing KL divergence between posterior distributions:

$$\min_q \mathrm{KL}(q_u(\theta) \| p(\theta | D_r))$$

where $q_u(\theta)$ represents the approximate posterior after unlearning.

Variational Formulation: Equivalently minimizes an Evidence Upper Bound (EUBO):

$$\mathcal{U} = \mathbb{E}_{q_u}[\log p(D_f | \theta)] - \mathrm{KL}(q_u(\theta) \| p(\theta | D))$$

Scope: Particularly effective for Bayesian neural networks and Gaussian processes. Requires modeling complete posterior distributions, limiting scalability to massive models.

3.2.1.7 PUMA: Performance Unchanged Model Augmentation

PUMA reweights remaining data samples to minimize utility degradation during unlearning:

$$\theta_u = \underset{\theta}{\text{argmin}} L(w_r \odot D_r, \theta)$$

where w_r represents learned importance weights for retained samples.

Key Innovation: Rather than directly modifying parameters, PUMA learns reweighting schemes approximating unlearning while maintaining model performance. This approach connects unlearning to importance-weighted learning objectives.

3.2.2 Meta-Learning and Optimization Approaches

3.2.2.1 Learning to Unlearn (LTU): Gradient Harmonization

LTU frames unlearning as a meta-learning problem, optimizing how to best update parameters for dual objectives: forgetting target data while remembering retain data.

Gradient Conflict Resolution: When optimizing competing objectives, conflicting gradients arise:

$$\cos(\nabla_f, \nabla_r) < 0 \text{ indicates conflicting directions}$$

LTU resolves conflicts through gradient projection:

$$\nabla_f^{'} = \nabla_f - \frac{\nabla_f \cdot \nabla_r}{\|\nabla_r\|^2} \nabla_r$$

Meta-Optimization: Employs bilevel optimization with gradient feedback—inner loop computes unlearning and remembering gradients, outer loop harmonizes conflicting components, combined update: $\nabla = \nabla_r + \nabla_f^{'}$.

Experimental Performance: On CIFAR-10 with ResNet-18 removing 10% of data, LTU achieves 4.37% unlearning accuracy gap (versus retrain baseline) while using only 30% of retained data, significantly outperforming methods requiring complete retain set access.

3.2.2.2 SFTC: Selective Fine-Tuning and Targeted Confusion

SFTC combines student–teacher fine-tuning with sophisticated confusion mechanisms:

Architecture: Uses original model as teacher for retained data; employs biased random generator for forgotten data; trains student to distinguish between retain classes while becoming uncertain about forget classes.

Targeted Confusion Innovation: Unlike previous random confusion approaches, SFTC employs biased random output generation governed by confusion fraction parameter β. When $\beta = 1$, outputs are completely random; when $\beta = 0$, outputs are biased toward correct labels with reduced confidence.

Empirical Results: On CIFAR-10 with ResNet-18, SFTC achieves 0.96% accuracy error on forget set versus retrained model (compared to 0.89% for simple fine-tuning baseline), with superior performance on unlearning metrics.

3.2.2.3 Zero-Shot Machine Unlearning

Addresses extreme deployment scenarios where no training data remains accessible after model deployment.

Error Minimization-Maximization: Generates synthetic "anti-samples" simultaneously:

- Maximizing error on forget classes (error-maximizing noise)
- Minimizing error on retain classes (error-minimizing noise)

Gated Knowledge Transfer: Uses generator networks with band-pass filters for selective knowledge transfer—generators produce pseudo-data points; filters ensure only retain-class knowledge transfers to student model.

Practical Advantage: Operates entirely post-deployment without original training data access. Critical for models deployed as services without internal data access.

3.2.2.4 Few-Shot Machine Unlearning

Extends unlearning when only partial forgotten data is available.

Model Inversion: Learns to invert model predictions to recover data distributions:

$$\min_{\theta} \mathrm{KL}(p_{recovered}(\cdot) \| p_{true}(\cdot))$$

Application: Useful when only extracted or partially recovered forget samples exist through membership inference attacks or user deletion requests.

3.2.3 *Knowledge Distillation Methods*

3.2.3.1 UNDIAL: Unlearning via Self-Distillation

UNDIAL modifies model output logits to encourage forgetting:

$$\mathcal{L} = \mathrm{KL}(\sigma(\mathrm{logits}_{\mathrm{adjusted}}) \| \sigma(\mathrm{logits}_{\mathrm{original}}))$$

where adjusted logits suppress probabilities for forgotten classes:

$$\mathrm{logits}_{\mathrm{adjusted}}(c) = \begin{cases} -\infty & \text{if } c \in C_f \\ \mathrm{logits}(c) & \text{otherwise} \end{cases}$$

Implementation: Simple to deploy on pretrained models; applicable to models without retraining from scratch.

3.2.3.2 RKLU: Reverse KL Divergence Knowledge Distillation

Employs reverse KL divergence emphasizing high-probability regions:

$$\mathrm{KL}(p_{\mathrm{teacher}} \| p_{\mathrm{student}}) = \mathbb{E}_{p_{\mathrm{teacher}}}[\log p_{\mathrm{teacher}} / p_{\mathrm{student}}]$$

Advantage: Naturally avoids probability mass on forgotten classes by focusing on teacher's high-probability regions.

3.2.3.3 ULD: Unlearning from Logit Difference

Employs auxiliary language models to generate reference outputs, learning differences:

$$\theta_u = \operatorname*{argmin}_{\theta} \| M_\theta(D_f) - M_{\mathrm{reference}}(D_f) \|$$

Mechanism: Assistant LLM generates baseline predictions; target model learns to match baselines (indistinguishable from random performance on forget data).

3.2.4 Neural Architecture Methods: Sparsity-Aware Unlearning

Critical Finding: Model sparsity significantly reduces the gap between approximate and exact unlearning.

Theoretical Analysis: The unlearning error for gradient ascent on sparse models:

$$\mathcal{E}_m = O\left(\frac{2\eta \|t\|_0^2}{\sigma_m^2}\right)$$

improves as sparsity increases (fewer active dimensions). Empirical evidence demonstrates 51% unlearning efficacy improvement and 8% MIA-efficacy improvement when models reach 95% sparsity.

Two-Step Approach:

1. Apply magnitude pruning (OMP) or training-free pruning (SynFlow)
2. Apply standard unlearning algorithm to sparse model

Sparsity-Aware Regularization: Incorporates L1 regularization directly:

$$\theta_u = \underset{\theta}{\text{argmin}} \mathcal{L}_u(\theta | D_o, D_r) + \lambda_s \|\theta\|_1$$

with sparsity-decay scheduler: $\lambda_s(t) = \lambda_0 \cdot (1 - t/T)^2$.

Transfer Learning Application: Removes harmful ImageNet classes before transfer learning, achieving $2 \times$ speedup over retraining while improving downstream task accuracy.

3.2.5 Specialized Techniques

3.2.5.1 In-Context Unlearning: Inference-Time Intervention

Rather than modifying parameters, manipulates prompts and context at inference time:

Mechanism: Prepend context priming models to generate generic responses for forgotten topics. Example: "You don't know about X. Respond with 'I cannot help with this.'".

Advantages: Black-box compatible; requires no model parameter access; applicable to proprietary LLM APIs.

Limitations: Operates through behavioral steering rather than knowledge removal; vulnerable to prompt injection and jailbreaking attempts.

3.2.5.2 IFRU: Influence Function-Based Recommendation Unlearning

Applies influence functions to recommendation systems:

$$\text{influence}(u, i) = \nabla_\theta L(u, i) \cdot H^{-1} \nabla_\theta L(\text{all data})$$

Efficiently identifies and removes user-item interactions without full retraining.

3.2.5.3 Scrub-And-Learn: Category-Aware Weight Modification

Applies category-specific modifications using submaximum one-hot encoding:

$$\text{target}_{\text{scrambled}}(c) = \begin{cases} 0 \text{ if } c \in C_f \\ 1 \text{ otherwise} \end{cases}$$

Guides model weights toward this target distribution while preserving other weights.

3.2.6 *Advanced Representation Methods*

3.2.6.1 Contrastive Unlearning: Push–Pull Embedding Reshaping

Operates in embedding space through:

1. **Pushing** forgotten sample embeddings away from original class clusters
2. **Pulling** them toward test set embeddings (random/uninformative)

Loss Function:

$$\mathcal{L} = -\log \frac{\exp(\text{sim}(\mathbf{z}_u, \mathbf{z}_t)/\tau)}{\sum_i \exp(\text{sim}(\mathbf{z}_u, \mathbf{z}_i)/\tau)}$$

Termination: Achieves unlearning when accuracy on forgotten class approaches zero.

Comprehensive Results: Generalizes across architectures (CNNs, Vision Transformers, CLIP); achieves 15 passes compared to 40 + for gradient ascent; superior utility preservation (81.86% versus 67.64% for gradient ascent on ResNet-18).

3.2.6.2 Localization-Informed Unlearning: Interpretability-Guided Updates

Identifies critical parameters through:

- **Causal Tracing**: Identifies layers most responsible for specific outputs
- **Saliency Analysis**: Computes gradient-based importance scores
- **Neuron Activation Analysis**: Finds neurons responding to forgotten concepts
- **Fisher Importance**: $\text{importance}_j = \mathbb{E}[\|\nabla_{\theta_j} L(D_f)\|^2]$

Targeted Update: Modifies only high-importance parameters, reducing computational cost and side effects.

Table 3.1 Comparison of different approximate unlearning techniques

Method	Efficacy	Utility	Speed	Requirements	Data access
Amnesiac	Excellent	Excellent	Very fast	Gradient storage	Full
Fisher	Good	Good	Slow	FIM computation	Full
Fine-tuning	Moderate	Excellent	Fast	Retain data	Partial
Gradient ascent	Very good	Moderate	Moderate	Careful tuning	Partial
Contrastive	Excellent	Very good	Fast	Embeddings only	None
LTU	Excellent	Excellent	Moderate	Meta-training	30% Retain
SFTC	Excellent	Very good	Fast	Random generator	Partial
Zero-Shot	Good	Good	Moderate	None	None
Sparsity-aware	Excellent	Very good	Very fast	Pruned model	Partial

3.3 Comparative Analysis and Deployment Considerations

3.3.1 Performance Trade-Off Matrix

Comprehensive comparison across key dimensions is shown in Table 3.1.

3.3.2 Selection Criteria for Practitioners

Small Forget Sets (<1% data): Amnesiac or Forsaken provide precise removal with minimal collateral damage. Gradient ascent becomes problematic due to noisy gradient estimates.

Large-Scale Models (Billions of parameters): Gradient-based methods with localization (identifying important parameters) reduce computational burden. Contrastive unlearning provides practical efficiency.

Privacy-Critical Applications: Contrastive unlearning demonstrates robust utility preservation alongside effective privacy protection. Sparsity-aware methods provide theoretical guarantees.

Post-Deployment Scenarios: Zero-shot or in-context unlearning operate without original data access. Critical for deployed models without data retention policies.

Federated Learning: FedEraser or FedUnlearn handle distributed data and privacy constraints. Parameter aggregation protocols prevent central data leakage.

Rapid Deployment Requirements: Sparsity-aware unlearning and contrastive methods achieve sub-second unlearning times for moderate forget sets.

3.3.3 Emerging Challenges and Future Directions

Theoretical Foundations: Current approximation bounds remain loose, providing limited guidance for practitioners. Tighter error analysis could inform algorithm selection.

Evaluation Methodology: Membership inference attacks provide practical privacy assessment but remain computationally expensive. New evaluation metrics independent of retrain-from-scratch oracles are needed.

Large Language Model Unlearning: Most algorithms were developed and evaluated on computer vision tasks. NLP domain specifics (tokenization, sequence length, attention mechanisms) require novel approaches.

Multi-Modal Unlearning: Emerging vision-language models require unlearning across modalities simultaneously. Current methods assume single modality inputs.

Certified Unlearning: Differentially-private approaches provide formal privacy guarantees but at significant utility cost. Balancing theoretical guarantees with practical performance remains open.

3.4 Conclusion

Approximate machine unlearning represents a pragmatic response to regulatory, privacy, and safety imperatives in deployed machine learning systems. Rather than seeking universally superior methods, the field has matured to recognize that different algorithms optimize distinct dimensions: computational efficiency, unlearning completeness, utility preservation, and scalability.

The progression from parameter-modification methods through meta-learning to representation-based approaches demonstrates increasing sophistication in addressing the fundamental challenge. Early approaches directly modified weights; contemporary methods leverage meta-optimization, contrastive learning, and architectural properties to achieve superior performance across multiple metrics simultaneously.

Empirical evidence from large-scale experiments demonstrates that contextual factors profoundly influence algorithm selection. Model architecture (dense versus sparse), dataset characteristics (balanced versus imbalanced), and deployment constraints (data access, computational resources) all significantly impact which unlearning algorithm delivers optimal results.

The field faces critical open problems: developing practical methods for large language models, creating evaluation metrics that avoid expensive retrain-from-scratch baselines, and balancing formal privacy guarantees with utility preservation. As regulatory pressure intensifies and model deployment becomes ubiquitous, approximate unlearning will likely become essential infrastructure in responsible machine learning systems.

Chapter 4
Machine Unlearning for Different Model Architectures

Abstract This chapter offers a thorough investigation into architecture-specific Machine Unlearning (MU) algorithms, illustrating that effective forgetting necessitates moving beyond model-agnostic techniques to utilize the structural characteristics of various ML models. We start by looking at the obstacles and emerging solutions for unlearning in Deep Neural Networks (DNNs), assessing key approximate methods such as the SAFE algorithm and the Amnesiac approach, which emphasize targeted weight adjustments and gradient alterations. The chapter then transitions to relational and sequential data models, outlining unlearning strategies for Graph Neural Networks (GNNs) and specifically examining the GraphEraser framework. It also investigates unlearning in practical applications like recommender systems, focusing on methods like RecEraser and MUSO that tackle the issue of data deletion in dynamic, high-dimensional user-item interaction contexts. Additionally, the chapter addresses the crucial area of Natural Language Processing (NLP), providing an analysis of techniques designed for Large Language Models (LLMs), including the S3T method for effectively handling concept and knowledge removal at scale. Lastly, the chapter examines decentralized unlearning within Federated Learning settings, contrasting efficient partitioning strategies such as Federated SISA (Sharded, Isolated, Sliced, Aggregated) with advanced, security-oriented protocols like Knot and SecureCut.

4.1 Introduction

Model Architecture-Specific Unlearning refers to techniques that modify the model's underlying structure or components or architecture, or parameters to remove data influence. The unlearning mechanism is explicitly tailored to the operational dynamics and components of a particular neural network architecture. By targeting specific layers, nodes, filters, or attention mechanisms responsible for encoding the "forgetting" data, the techniques aim to achieve high unlearning efficacy with drastically reduced computational cost compared to full model-agnostic methods. For Convolutional Neural Networks (CNNs), unlearning often involves strategies to mitigate the influence of specific images or classes, like using neural masking (e.g.,

R. Gupta et al., *Machine Unlearning: Concepts, Techniques and Applications*,
Studies in Computational Intelligence 1270,
https://doi.org/10.1007/978-3-032-18447-4_4

to unlearn the "plane" class from a VGG16 model trained on the CIFAR-10 dataset) or knowledge distillation, where a "clean" model transfers its knowledge to a student model to forget the target data. For Large Language Models (LLMs), which often use the Transformer architecture, methods like LoRA (Low-Rank Adaptation) fine-tuning and gradient ascent are utilized to efficiently reverse the impact of specific text data or sensitive knowledge without full retraining. The challenges vary by architecture; for CNNs, it's the widely distributed nature of features across layers, while for LLMs, it's the sheer size of the model. This chapter explores how different architectures—from simple Multilayer Perceptrons to complex Transformers—can be leveraged to perform precise data removal.

4.2 Unlearning in Deep Neural Networks

Unlearning in Deep Neural Networks (DNNs) presents a significant challenge, moving beyond simple data deletion to efficiently and provably remove the influence of specific training data from a complex, non-convex model without the time and resource expenditure of full retraining. It operates across several architectural levels, from individual parameters to entire layers, addressing the challenge of removing the influence of specific training data without complete retraining (refer Fig. 4.1). At the most fundamental weight parameter level, unlearning is achieved through precise modifications like gradient-based adjustments, which effectively subtract the forgotten data's accumulated influence, or through weight pruning, which sets redundant or non-essential parameters to zero, thereby eliminating their ability to carry residual information. Moving up the hierarchy, layer-wise unlearning employs strategies like selective layer retraining, which focuses computationally expensive gradient updates exclusively on the higher, task-specific layers most affected by the forgotten data, leaving lower, general feature-extracting layers frozen. Beyond parameter changes, unlearning can target the hidden layer activations by suppressing the internal representations strongly associated with forgotten samples, a process influenced by the network's activation function choice (e.g., ReLU's sparsity aids pruning). Finally, achieving these goals for large-scale DNNs faces a computational bottleneck, necessitating the use of approximations like the Fisher Information Matrix (FIM) or selective parameter computation instead of the infeasible, full Hessian matrix operations required for exact unlearning. Effective implementation ultimately requires the coordinated integration of these architectural mechanisms, balancing the trade-off between the computational cost of exact forgetting and the efficiency of scalable approximate methods.

This field has seen the emergence of advanced algorithmic approaches that tackle the core problem—balancing the need for high unlearning quality (ideally matching a model retrained from scratch) with computational efficiency. Algorithms like SAFE (Synergy Aware Forgetting Ensemble) (Dukler et al. 2023) and LegoNet achieve this balance through architectural modularity and strategic data sharding, designing the network structure itself to isolate the impact of individual data subsets (Yu et

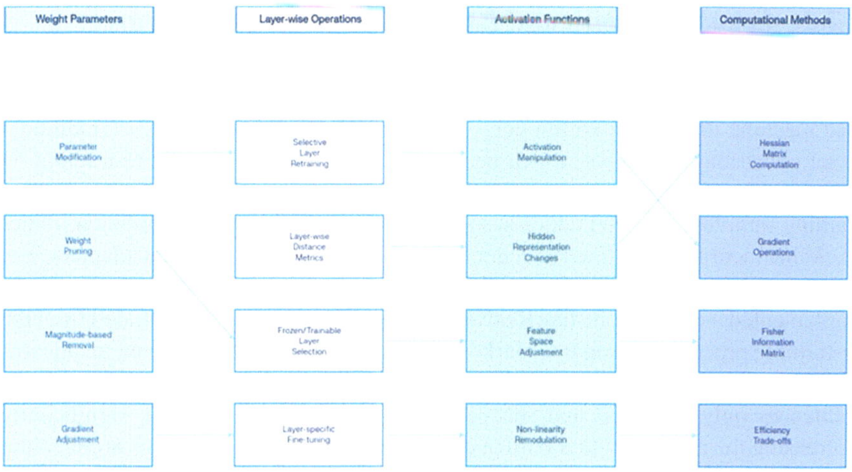

Fig. 4.1 Machine unlearning in deep neural networks

al. 2022). In contrast, techniques such as Amnesiac Unlearning (including variants like MaGA) pursue efficiency via approximate unlearning, employing methods like targeted parameter perturbation or strategic label noise injection to quickly erase the forgotten data's influence with bounded impact on the model's overall utility (Graves et al. 2021).

4.2.1 *Synergy Aware Forgetting Ensemble (SAFE)*

SAFE algorithm is designed to enhance the traditional SISA paradigm by introducing a novel approach of using a shard graph (Dukler et al. 2023). This graph allows the individual models to incorporate limited, structured information from other shards during training. By doing so, SAFE leverages the synergistic information among various shards that would otherwise be lost in a naive sharding (SISA) approach. At its core, SAFE operates on a pre-trained deep learning model. This backbone model is typically a large, complex neural network (like a convolutional neural network for images or a Transformer for text) that has been trained on a massive dataset. Thus, the input to SAFE is a training dataset partitioned into multiple shards with associated shard-level relationships. The output is an ensemble of adapter-based models, each specialized yet synergistically aware. The steps include constructing the shard graph, training adapters with shared backbone features and neighbor-aware inputs, and retraining selectively upon unlearning requests.

Upon a deletion request, only the adapter associated with the affected shard and any dependent adapters require retraining. The backbone is frozen during the unlearning process, meaning its millions or even billions of parameters are not changed.

4.2.2 LegoNet: A Fast and Exact Unlearning Architecture

LegoNet is an innovative network architecture tailored for achieving swift and precise machine unlearning within deep neural networks (Yu et al. 2022). Traditional exact unlearning strategies, which necessitate comprehensive model re-training on retained data, are often inefficient due to the vast number of model parameters and training samples. LegoNet addresses this efficiency challenge by adopting a framework that incorporates a fixed encoder along with multiple independent adapters. The fixed encoder, which encompasses the majority of the model parameters (typically implemented as a backbone like ResNet or Vision Transformer), is excluded from the re-training process, leading to a marked reduction in parameters involved in unlearning. Moreover, the adapter activation mechanism ensures that each training sample influences only a limited and specific number of adapters ($k \ll n$), significantly decreasing the computational burden during the unlearning phase. This architecture enables LegoNet to achieve exact unlearning with enhanced speed while preserving satisfactory model performance.

4.2.3 Amnesiac

The Amnesiac framework provides an approximate machine unlearning solution designed to remove the influence of specific training data from deep neural networks, a crucial task given that models are vulnerable to privacy risks like model inversion and membership inference attacks (Graves et al. 2021). Recognizing that simple data removal is insufficient to prevent information leakage, Amnesiac proposes two highly efficient methods that ensure the post-unlearning model is both secure against such attacks and maintains high predictive efficacy. The first method, Amnesiac Retraining, leverages the phenomenon of catastrophic interference by retraining the model for a short period on a carefully selected subset of the retained data, causing targeted forgetting of the deleted information. The second, Amnesiac Transfer Learning, accelerates the unlearning process by freezing the early, general feature layers and only retraining the final layers on the retained data, isolating the necessary parameter updates and preserving the model's fundamental knowledge.

Difference Between SISA, SAFE, LegoNet, and Amnesiac

The core distinction is that SISA trains s independent, full sub-models, each on a small, disjoint shard of data. When a sample is unlearned, only the single sub-model corresponding to its shard is re-trained. SAFE improves upon SISA by introducing a Shard Graph that allows limited, optimized information flow between the sharded sub-models or adapters during training. This controlled connectivity mitigates the accuracy degradation seen in SISA, allowing SAFE to achieve a better balance of fast unlearning (by only retraining components) and high final model performance. Further, the SISA sub-model includes an entire encoder and adapter, meaning the number of retrainable parameters remains huge for each unlearning task while LegoNet focuses on reducing both parameters and samples. LegoNet uses a single, fixed encoder shared across n adapters, removing the vast majority of parameters (the encoder) from the re-training scope entirely. For unlearning, LegoNet only re-trains **k** small adapters, resulting in significantly fewer parameters to update (e.g., thousands of times less). Amnesiac focuses on what to undo (parameter updates) in a single step, trading storage for speed, whereas LegoNet, SAFE, and SISA focus on what to retrain (a small component), trading a small amount of retraining time for a massive gain in efficiency over full retraining.

Comparison of Experimental Results of SAFE, Legonet and Amnesiac

Datasets The three unlearning algorithms utilize image classification datasets with varying levels of complexity. LegoNet is predominantly evaluated on common vision benchmarks, specifically CIFAR-10 and CIFAR-100. CIFAR-10 contains 60,000 tiny (32×32 pixel) color images across 10 classes, while CIFAR-100 is more complex, featuring the same number of images distributed across 100 classes, grouped into 20 superclasses, making it a more challenging benchmark for general object recognition. Amnesiac Unlearning is evaluated across a diverse range of tasks, often employing standard vision benchmarks like MNIST (a simple dataset of 28×28 grayscale handwritten digits) and CIFAR datasets, but also focusing on scenarios where privacy leakage is measured using metrics like Membership Inference Attacks (MIA). Finally, the SAFE method specializes in fine-grained computer vision datasets and large models. Fine-grained datasets, unlike general classification datasets, require distinguishing between subordinate classes (e.g., different species of birds or types of cars) and are characterized by high inter-class similarity and typically involve larger models for accurate classification. These varied datasets ensure the unlearning methods are tested against simple, complex, and highly specialized data distributions.

Results The experimental comparison of three prominent machine unlearning algorithms—SAFE, Amnesiac, and LegoNet—reveals distinct trade-offs between unlearning efficiency, effectiveness, and retained model utility. LegoNet, primarily evaluated on image classification datasets like CIFAR-10 and CIFAR-100, focuses on exact unlearning by employing a fixed encoder with multiple adapters; its results demonstrate significant efficiency improvements over full retraining, minimizing the parameters that need updating during a forget request, yet its utility can sometimes be lower than the 'gold' retrained model due to the fixed backbone. In contrast, Amnesiac Unlearning is evaluated across diverse tasks, often using standard vision benchmarks (e.g., MNIST, CIFAR) but also incorporating privacy-centric metrics like Membership Inference Attack (MIA) success rates; this approach emphasizes efficient approximate unlearning, showing high efficacy in mitigating privacy leakage, though its unlearning mechanism may not carry the same theoretical guarantees as exact methods. Finally, the SAFE (Synergy Aware Forgetting Ensemble) method distinguishes itself by specializing in fine-grained computer vision datasets and large models, leveraging a shard graph structure and lightweight adapters; experiments show SAFE reduces the cost of forgetting by an order of magnitude compared to similar sharding techniques, while crucially maintaining high model accuracy by optimizing synergistic information between data shards. Overall, while LegoNet prioritizes exactness and SAFE prioritizes the efficiency-utility trade-off in complex computer vision tasks, Amnesiac focuses broadly on ensuring privacy protection through approximate, efficient forgetting across varied datasets.

4.3 Modular Unlearning in Over-Parameterized Regimes

The primary impediment to the practical application of Machine Unlearning (MU) in the context of modern Deep Neural Networks (DNNs)—particularly large language models—is the intractable computational expense associated with modifying models possessing billions of parameters. MUSO (Machine Unlearning with Subspace Optimization) constitutes a critical methodology designed to mitigate this issue (Yang et al. 2024). This approach leverages the insight that the requisite weight adjustments for data erasure are not uniformly distributed but are instead concentrated along a limited number of principal directions within the high-dimensional weight space. Consequently, instead of updating the entire parameter set, MUSO restricts the unlearning operation to a specialized, computationally efficient subspace.

4.3.1 Machine Unlearning with Subspace Optimization (MUSO)

Context and Motivation MUSO is specifically designed to handle the massive size (the "over-parametrized" nature) of modern Deep Learning models, such as large language models (LLMs) or complex deep vision networks. It operates by performing highly efficient, targeted updates to the model's weight matrix and uses optimization techniques tailored for their high dimensionality. The process is structured into three essential, sequential phases:

1. **Estimating the Ideal Perturbation (The Target)** The initial phase involves calculating the precise, complete weight perturbation required to neutralize the influence of the target data. This yields the ideal "change vector," which serves as the ultimate target for the unlearning procedure. This estimation is typically derived through advanced influence analysis techniques.
2. **Identifying the Low-Dimensional Subspace (The Manifold)** The core innovation of MUSO is the identification and construction of a low-dimensional subspace capable of accurately encompassing the ideal change vector. This subspace is spanned by a small set of basis vectors. Should a model contain D parameters, the effective dimensionality of the unlearning task is reduced to k, where k is substantially smaller than D. The derivation of this subspace is frequently accomplished by analyzing latent structures, such as gradient patterns, preserved within the model's architecture.

3. **Subspace Optimization and Application** Subsequently, the estimated ideal perturbation is projected onto this restricted subspace. The unlearning problem is thereby reformulated into an efficient optimization task: determining the optimal set of coefficients that define the necessary adjustment within this sparse subspace. The final unlearned model is constructed by adding this subspace-optimized perturbation to the original model's weights, rendering the computational overhead negligible compared to full model retraining.

MUSO is formally categorized as a High-Fidelity Approximate Unlearning method. It does not meet the criteria for certified mathematical exactness relative to a model retrained from initialization, as the projection onto a constrained subspace inherently introduces a minimal approximation error. Nevertheless, its practical efficacy is supported by robust empirical validation, demonstrating successful removal of data influence while maintaining model utility at unprecedented speeds.

The experimental evaluation of MUSO demonstrates substantial performance benefits, particularly when applied to large, over-parametrized Deep Neural Networks. Results typically show that the method reduces the computational cost of unlearning by orders of magnitude compared to standard full retraining. Crucially, this efficiency is achieved with high fidelity, meaning the utility (accuracy) of the unlearned model on retained data closely matches the "gold standard" model trained only on that retained data. Furthermore, MUSO's efficacy in removing target data influence is confirmed by its ability to significantly mitigate privacy risks, achieving results empirically comparable to the ideal retrained model, thus establishing it as a highly scalable and practical solution for machine unlearning in industry.

4.4 Graph Neural Network Based Machine Unlearning

Machine unlearning (MU) in Graph Neural Networks presents unique challenges due to the interdependencies inherent in graph-structured data, unlike the relative independence of samples in image or tabular data. The core difficulty lies in the message-passing architecture of GNNs, where each node's representation is refined by aggregating information from its neighbors across multiple layers. When an unlearning request (e.g., deleting a node v) occurs, its influence cascades through multi-hop neighborhoods, meaning simply removing the data is insufficient; the residual "neighborhood influence" remains encoded in neighboring node representations.

Effective GNN unlearning requires Graph Property-Aware Approaches that address three main request types: Node Unlearning (removing a node and its edges), Edge Unlearning (removing specific relationships), and Attribute Unlearning (removing specific features). Methods like GraphEraser tackle node unlearning by employing balanced label propagation and clustering to partition the graph, preserving structural properties crucial for model utility. GNNDelete addresses edge unlearning by inserting localized, trainable transformation operators ($\mathbf{T}^{(l)}$) after GNN layers, ensuring that information flow is halted only within the affected multi-hop subgraph. Other approaches leverage Graph Influence Functions (GIF) to estimate

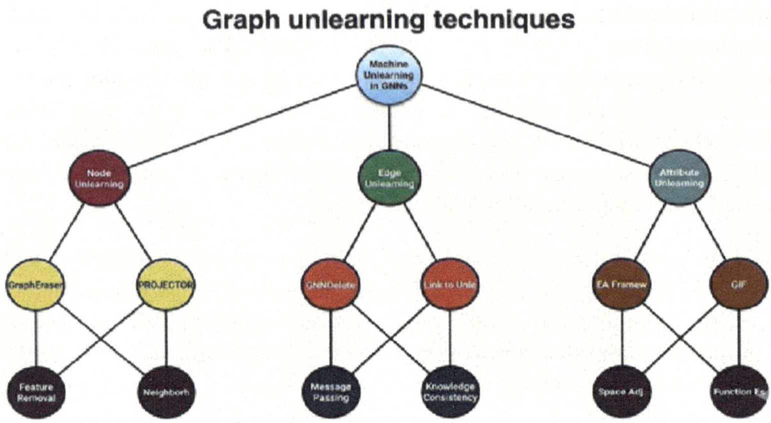

Fig. 4.2 Graph unlearning techniques

parameter changes while explicitly accounting for multi-hop neighborhood effects, or utilize Projection-Based Unlearning (PROJECTOR) to achieve exact unlearning for certain GNNs by projecting weight matrices onto null subspaces orthogonal to forgotten features. The approaches for graph unlearning are summarised in Fig. 4.2. Ultimately, successful GNN unlearning must move beyond simple parameter modification to explicitly model and preserve graph topology while systematically eliminating information propagated via the message-passing mechanism.

Subsequent sections explain some of the prominent algorithms in GNNs.

4.4.1 GNNDelete

GNNDelete is a general, model-agnostic approach designed to handle the deletion of nodes, edges, or node features (Cheng et al. 2023). It introduces a novel layer-wise deletion operator that learns small, lightweight weight matrices instead of retraining the entire model. Its optimization objective enforces two key properties: Deleted Edge Consistency (making deleted elements look like non-existent ones) and Neighborhood Influence (preserving knowledge of the remaining graph structure).

4.4.2 GraphEraser

GraphEraser is a partitioning and ensemble-Based Unlearning algorithm based on SISA framework (Zhang et al. 2022).

Core Mechanism The fundamental idea of GraphEraser is to convert the expensive task of retraining a single large GNN into the much cheaper task of retraining only a small subset of constituent models. It achieves this through the following steps:

1. **Graph Partitioning:** The full training graph $\mathcal{G}$ is partitioned into K disjoint subgraphs (shards), $\mathcal{G} = \{\mathcal{G}_1, \mathcal{G}_2, \ldots, \mathcal{G}_K\}$. GraphEraser proposes two novel partitioning strategies to minimize structural damage: one based purely on community detection (preserving structural information) and one that considers both graph structure and node features (clustering in the embedding space).
2. **Shard Model Training:** A separate, independent GNN model, $\mathcal{M}_i$, is trained on each shard $\mathcal{G}_i$. The predictions of these K models are then aggregated to produce the final prediction of the ensemble GNN model $\mathcal{M}$.
3. **Unlearning Execution:** When an unlearning request for an element (node, edge, or feature) is received, GraphEraser first identifies which shard $\mathcal{G}_i$ contains the forgotten element. Only the corresponding model $\mathcal{M}_i$ is retrained from scratch (or fine-tuned) on the modified shard $\mathcal{G}_i'$. The other $K - 1$ models remain untouched.
4. **Prediction Aggregation:** The final prediction after unlearning is generated by aggregating the outputs of the updated model $\mathcal{M}_i'$ and the $K - 1$ retained models.

GraphEraser offers a significant boost in efficiency, as the time complexity of unlearning is reduced from that of retraining the entire model on $\mathcal{G}$ to retraining a much smaller model on a shard $\mathcal{G}_i$. The main limitation stems from the partitioning step. Since GNNs rely heavily on neighborhood information and message passing across the entire graph, splitting the graph into truly disjoint shards inevitably damages the overall structural integrity. This can lead to a drop in the utility or performance of the unlearned model compared to a model fully retrained on the complete, retained graph.

4.4.3 Computational and Logical Trade-Offs in Graph Unlearning

Machine unlearning in GNNs necessitates careful navigation among competing objectives inherent to graph properties which are listed below:

1. **Scalability versus accuracy:** Partition-based methods like GraphEraser offer excellent scalability by enabling parallel unlearning through partial retraining, but sub-model degradation from disconnected subgraphs impacts overall model utility.
2. **Flexibility versus efficiency:** Operator-based methods like GNNDelete provide flexibility to handle diverse unlearning requests with theoretical properties but demand iterative optimization.
3. **Exactness versus practicality:** Projection-based approaches guarantee exactness for specific GNN architectures but may not generalize across diverse GNN variants.

The underlying computational constraint stems from GNN architectures' inherent multi-hop dependencies. Unlike standard models where computational complexity scales with parameter count, GNN unlearning additionally depends on graph diameter (maximum shortest path between nodes) and average degree, as these metrics determine how extensively unlearning effects propagate through message passing. Effective machine unlearning in Graph Neural Networks demands architectural understanding far deeper than traditional deep learning, necessitating explicit modeling of graph properties including node dependencies, edge connectivity, multi-hop information propagation, and structural integrity. Successful approaches move beyond parameter-centric unlearning toward property-aware mechanisms that exploit and preserve graph topology while systematically eliminating forgotten information's influence through message-passing mechanisms. As graph-based applications increasingly demand privacy-preserving data management, continued advancement in property-aware unlearning approaches remains critical for trustworthy graph neural network deployment.

4.5 Machine Unlearning for Recommendation Systems

Machine Unlearning (MUL) in recommendation systems (RecSys) presents unique challenges compared to domains like image classification, primarily due to the collaborative and interconnected nature of user-item interaction data. The removal of one user's interaction affects the embedding vectors of shared items and other users. This necessity for maintaining collaborative properties post-unlearning has spurred the development of specialized algorithms. The dominant technique, matrix factorization, decomposes the sparse user-item interaction matrix R into low-rank user embeddings U and item embeddings V. The core difficulty for unlearning stems from this interdependence: removing a user not only affects their personal embedding U_i but also subtly influences the item embeddings V learned from their interactions. Furthermore, recommendation data is typically extremely sparse (0.01–1% dense), meaning each observed interaction carries significant weight, creating complex parameter interdependencies. Consequently, general MU techniques like SISA, which rely on arbitrary data partitioning, severely degrade performance because they disrupt the essential collaborative information structure necessary for high-quality recommendations. Effective MU in this domain requires targeted adjustments to both user and item embedding matrices while carefully preserving the underlying collaborative patterns.

4.5.1 RecEraser: Partition-Aggregation for Recommendations

Recommender systems have become indispensable in applications such as e-commerce and social media, yet their reliance on historical user-item interactions raises privacy concerns. RecEraser addresses the problem of exact machine unlearning in recommendation tasks by ensuring that once a user requests data removal, the system can effectively and efficiently erase its influence without full retraining (Chen et al. 2022). RecEraser adopts a sharded submodel architecture inspired by the SISA framework but tailored for collaborative filtering. The model training dataset, representing user-item interactions, is divided into K balanced shards $\{S_1, S_2, ..., S_K\}$ via novel partitioning algorithms: User-based (UBP), Item-based (IBP), and Interaction-based Balanced Partitioning (InBP). Each shard S_i is used to train a submodel M_i. Unlike random splits, these methods preserve collaborative patterns by leveraging similarity measures in pretrained embedding spaces. To produce a final prediction, RecEraser introduces an attention-based adaptive aggregation method that computes user- and item-specific weights for each submodel. The aggregated output is optimized using base model loss and regularization coefficient. In the event of an unlearning request, only the affected submodel and the aggregation component need retraining, substantially reducing computational cost compared to full retraining. RecEraser thus represents a scalable, performant, and legally compliant solution for exact unlearning in recommendation systems.

Experimental Results and Analysis Experiments show that RecEraser achieves unlearning completeness while generally outperforming state-of-the-art baselines like SISA and GraphEraser in terms of model utility (e.g., HR@K, NDCG@K) across various datasets and base models (BPR, LightGCN). Its core strength is the custom data partition and dynamic aggregation, which effectively preserves collaborative information, thus mitigating the utility drop typical of ensemble methods. However, its weakness lies in the fact that, as an ensemble approach, its unlearning efficiency is still limited by the need to retrain multiple sub-models.

4.5.2 Matrix Correction Approaches: IMCorrect Framework

The IMCorrect framework offers an efficient, non-retraining solution for unlearning in recommendation systems by using matrix decomposition theory. It treats the model output as an interaction matrix multiplied by a mapping matrix (**W**) and, upon an unlearning request, directly computes the required correction to the mapping matrix using closed-form calculations. This approach avoids costly iterative retraining, achieving unlearning in constant time relative to the dataset size, and provides the added benefit of interpretability into how specific interactions influence predictions.

4.5.3 Sequential Training and Collaborative Preservation: LASER Framework

The LASER (Learning and unlearning for Erasable Recommendation) framework uses sequential training on data shards, unlike parallel partition methods, to facilitate unlearning while preserving collaborative information. It partitions data using hypergraph-based user clustering to group similar users, maximizing collaborative patterns within each shard. By training models sequentially on these shards, following a curriculum learning order, knowledge progressively accumulates, ensuring the final model effectively incorporates and preserves comprehensive user-item relationships across the entire dataset.

4.5.4 Influence Function-Based Approaches

Influence function methods adapted for recommendation systems identify interactions that most significantly affect model predictions using second-order optimization techniques. These methods calculate influence scores to rank training interactions and selectively update only the relevant user and item embeddings. While efficient as they avoid full retraining, they are approximate because they often update only a subset of parameters, meaning residual information about the removed interactions might remain encoded in the unmodified embedding dimensions.

4.5.5 Computational Considerations and Model Utility Preservation

Machine unlearning in recommendation systems necessitates carefully managing three competing objectives. Completeness requires thorough removal of target data influence from learned embeddings. Efficiency demands rapid unlearning without proportional costs to dataset size. Utility preservation requires maintaining recommendation quality for remaining interactions and users. The choice among approaches involves fundamental trade-offs. RecEraser trades efficiency for near-perfect collaborative information preservation through careful sharding; IMCorrect achieves exceptional efficiency through matrix-based corrections; LASER improves utility through sequential processing at computational cost; influence-based methods sacrifice completeness for parameter selectivity.

4.5.6 Federated Recommendation Unlearning

Federated recommendation systems, where users maintain local data without centralizing it to a server, require specialized unlearning approaches. Client-level unlearning becomes necessary when individual participants request deletion, requiring coordinated updates across distributed clients without centralizing data. These federated settings introduce additional complexity as global model aggregation mixes influences from multiple clients, making precise client-level removal particularly challenging. Machine unlearning in recommendation systems demands sophisticated understanding of collaborative filtering properties, embedding relationships, and interaction dependencies. Successful approaches move beyond generic data removal techniques toward recommendation-specific methods that explicitly preserve collaborative information while achieving efficient, complete unlearning. As privacy regulations increasingly mandate user deletion rights and recommendation systems become ubiquitous in e-commerce, content platforms, and personalization services, advancing recommendation unlearning techniques remains critical for building trustworthy, privacy-respecting systems.

4.6 Machine Unlearning in Natural Language Processing Models

Machine unlearning (MU) in NLP models presents unique challenges due to the autoregressive nature and the distributed encoding of knowledge across billions of parameters in transformer architectures. Unlike simpler models, knowledge in Large Language Models (LLMs) is embedded throughout the system, making straightforward removal difficult. Successful MU requires architectural specificity, focusing on the token-level encoding of knowledge, where specific tokens (like proper nouns or entities) function as semantic anchors carrying disproportionate information. By concentrating gradient-based updates on these high-importance tokens, rather than entire sequences, practitioners can achieve more targeted knowledge removal.

Common approaches include Gradient Ascent (GA), which maximizes prediction loss to reverse learned associations, though this risks catastrophic collapse. More stable alternatives like Negative Preference Optimization (NPO) frame unlearning as explicitly rejecting undesirable outputs. Furthermore, specialized methods target the architecture itself: Attention-shifting approaches directly manipulate attention weights to suppress fact-bearing tokens, while Representation Misdirection Unlearning (RMU) adjusts lower transformer layers to redirect internal token representations toward noise vectors, inducing selective amnesia.

Locate-then-unlearn approaches improve efficiency by identifying and targeting the specific parameters most critical for encoding forgotten information using attribution scores. Recent work also explores Subspace-Based Unlearning, which identifies and projects updates onto parameter subspaces irrelevant to the knowledge being forgotten. Evaluation remains challenging, as simple probability metrics are insufficient; adversarial evaluation using paraphrasing and jailbreak prompts is necessary to probe the robustness against residual knowledge. These complex methods must also contend with continual unlearning—managing sequential unlearning requests without catastrophic utility degradation.

Extensions of SISA for NLP: SISA-FC and SISA-A To address the limitations of the original SISA framework in NLP applications—particularly its inefficiency in handling large language models like BERT, two novel extensions are proposed: SISA-FC and SISA-A (Kumar et al. 2023). While SISA partitions data into disjoint shards and slices, storing full model checkpoints after each slice, this approach becomes impractical for NLP due to high storage and computation costs. SISA-FC mitigates this by fine-tuning only task-specific fully connected layers on top of a frozen language model, drastically reducing memory and retraining time. SISA-A goes further by using adapter modules within the transformer layers, updating only a small fraction (1–5%) of model parameters. This method achieves accuracy nearly equivalent to full fine-tuning while maintaining low memory and time footprints. A niche contribution of the paper is the strategic partitioning of data to minimize affected slices during unlearning, thus enhancing efficiency. The approaches are evaluated on GLUE tasks including SST-2, QQP, and MNLI, and demonstrate up to 100x speedups in retraining with minimal accuracy degradation. Unlike original SISA, these extensions are tailored for high-dimensional NLP models and offer more practical unlearning. However, they remain constrained to encoder-based architectures and show limited performance in few-shot settings.

4.7 Unlearning in LLM-Based Architecture

4.7.1 S3T: Sequence-Aware Sharded Sliced Training for Exact Machine Unlearning

S^3T, or Sequence-aware Sharded Sliced Training, is an exact unlearning framework developed to improve deletion capabilities in modern large language models (LLMs) and other sequential tasks (Chowdhury et al. 2025). While prior exact methods like SISA reduced re-training costs by partitioning data, S^3T directly addresses the major practical challenge of system downtime by incorporating Parameter-Efficient Fine-Tuning (PEFT) techniques, which significantly reduces the computational expense of re-training a component (Chowdhury et al. 2025).

Core Principles

- Sharding and Slicing: S^3T first employs Sharded training, partitioning the entire dataset into disjoint shards, each training an independent model component. It then uses Slicing, partitioning each component's parameters into a frozen base model and a small, re-trainable adapter, leveraging PEFT.

- Sequence-aware Grouping: For sequential data, a Sequence-aware grouping strategy ensures that related sequences or turns within a conversation are kept together in the same shard/component. This minimizes the complexity of deletion requests spanning multiple components and maximizes training integrity for the sequential task.
- Parameter-Efficient Fine-Tuning (PEFT): The core innovation is using PEFT, such as LoRA (Low-Rank Adaptation), to train only a minimal fraction of parameters (the adapters) for each component. This drastically reduces the computational load and time needed for re-training upon deletion.

Experimental Analysis The main strength of S^3T is its superior scalability and efficiency for exact unlearning, particularly in modern LLM-based production systems. By integrating PEFT, S^3T minimizes the retraining burden to a tiny subset of parameters, drastically reducing service downtime and cost associated with deletion requests. The Sequence-aware grouping is another strength, making the framework robust for handling complex sequential data. The primary weakness lies in the inherent trade-off of exact unlearning systems: developers must balance the budget for frequent offline re-training against the potential revenue loss from system halts during component updates. Additionally, the framework's performance heavily relies on the assumption that the chosen PEFT technique is expressive enough to achieve high accuracy when trained with only small, sharded amounts of data.

4.8 Federated Learning

Federated Learning (FL) is a decentralized machine learning approach where multiple devices (clients) collaboratively train a shared model without sharing raw data. Each client trains the model locally on its own data and sends only model updates to a central server, which aggregates them to update the global model. This preserves data privacy, reduces communication of sensitive information, and complies with regulations like GDPR. FL is widely used in applications like keyboard prediction, healthcare, and personalized services. Despite its benefits, FL faces challenges like data heterogeneity, communication overhead, and client reliability. Several techniques for machine unlearning in federated settings are explained below.

4.8.1 Sharded: Scalable Federated Unlearning via Isolated and Coded Sharding

The Scalable Federated Unlearning via Isolated and Coded Sharding framework is proposed to address the significant challenges of storage overhead and computational resources associated with providing the right to erasure in Federated Learning (FL) (Lin et al. 2024). While FL is a promising paradigm for collaborative model training with client privacy, the unlearning process typically requires many clients to participate in re-training, consuming substantial resources. This framework focuses on reducing both the number of affected clients and the server's storage requirements for model parameters, thereby making federated unlearning practical and scalable for real-world deployment. The method ensures that the removal of client-level data effect is achieved without affecting the performance of the final global model (Lin et al. 2024).

Core Principles

- Isolated Sharding: Clients are divided into multiple isolated shards across different training stages. This isolation prevents the data influence of a client in one shard/stage from propagating to clients in other shards, significantly limiting the number of clients that must re-train during an unlearning request.
- Coded Computing Mechanism: To mitigate the server's extensive storage overhead (which would typically store all sharded model parameters), a coded computing mechanism is employed. This mechanism compresses the model parameters across the clients and shards, allowing the server to efficiently store and reconstruct the necessary model components for aggregation and unlearning.

Experimental Results Experiments were conducted using image classification tasks on datasets like Fashion-MNIST and CIFAR-10, employing common FL algorithms like FedAvg and FedProx as baselines. The results demonstrated the scalability of the framework by showing that it successfully achieved federated unlearning while significantly reducing the storage overhead on the central server compared to traditional sharding methods. Crucially, the approach maintained high model accuracy for the final global model, with performance being comparable to the fully trained baseline. The isolated sharding proved effective in minimizing the number of clients involved in the re-training process, thereby reducing the overall wall-clock time required to fulfill an unlearning request.

4.8.2 KNOT: Asynchronous Federated Unlearning via Optimized Clustered Aggregation

KNOT is a novel clustered aggregation mechanism proposed to address the challenging problem of exact machine unlearning in Asynchronous Federated Learning (AFL) environments (Su and Li 2023). In AFL, client updates are aggregated at the server over multiple rounds, causing the effect of a single client's data to gradually permeate across many other clients. When a client requests its data to be erased, all these affected clients would traditionally need to participate in the re-training process, leading to prohibitive costs in terms of communication and wall-clock time. KNOT mitigates this challenge by modifying the aggregation process itself, localizing the influence of client data to a small subset of the system (Su and Li 2023).

Core Principles

- Clustered Aggregation Mechanism: KNOT divides clients into disjoint clusters ($\mathcal{C}_1, \ldots, \mathcal{C}_k$) before training begins. Crucially, the server performs model aggregation only within each cluster, preventing updates from one cluster's clients from directly influencing the model parameters of other clusters.
- Lexicographical Minimization: The assignment of clients to clusters is optimized by solving a lexicographical minimization problem. This optimization aims to minimize both the unlearning cost (limiting the number of affected clients) and the training time, ensuring the clustered aggregation maintains good model performance.
- Model Independence: By restricting aggregation within clusters, the base model trained by one cluster remains independent of models trained by other clusters with respect to the client-specific updates. This independence is key to localizing the unlearning task.

Experimental Results Experimental evaluation of KNOT was conducted on image classification tasks (CIFAR-10) and language modeling tasks (LEAF datasets) using various models like CNNs and LSTMs, comparing its performance against standard AFL algorithms like FedAsync (Su and Li 2023). The results demonstrated that KNOT successfully achieved exact federated unlearning while delivering a substantial reduction in unlearning cost with respect to the wall-clock time compared to existing methods. By limiting the re-training scope, the overhead required to satisfy an unlearning request was drastically reduced. The clustered aggregation strategy was shown to maintain a high level of accuracy, with the final model performance being comparable to or only slightly lower than the unlearnable FedAsync model, validating the effectiveness of the lexicographical minimization in optimizing the cluster assignments (Su and Li 2023).

4.8.3 SecureCut: Exact Machine Unlearning in Vertical Federated GBDTs

SECURECUT is a novel GBDT framework designed for Vertical Federated Learning (VFL) that efficiently enables machine unlearning in response to the right-to-be-forgotten mandate (Li et al. 2025). The challenge in VFL is that user data is fragmented, with different parties holding different feature subsets for the same set of samples, meaning data erasure requires removing specific features across all samples under privacy guarantees. Unlike re-training the entire tree structure, which is computationally prohibitive, SecureCut leverages the robust structure of GBDTs to allow for both instance unlearning (removing a row/sample) and feature unlearning (removing a column/feature set) without rebuilding the model from scratch, significantly reducing the cost of model updates (Li et al. 2025).

Core Principles

- Federated GBDT Structure: The framework is built upon a VFL-compatible GBDT where each party contributes features and the trees are built collaboratively, ensuring privacy protection during training through secure split finding.
- Structure-Agnostic Unlearning: SecureCut enables unlearning without modifying the established tree structure or the split points of the existing trees. This is achieved by adjusting the leaf values and instance weights only, which are computationally cheap operations.
- Instance/Feature Isolation: The algorithm defines precise mechanisms to isolate the influence of the deleted instance or feature subset on the residual errors and leaf value calculations across the entire GBDT ensemble.

Experimental Results Extensive experiments were conducted on popular datasets using various VFL settings, comparing SecureCut's performance and unlearning efficiency against state-of-the-art methods like DeltaBoost and HedgeCut. The results demonstrated that SecureCut achieved superior model utility (higher prediction performance) and forgetfulness (lower influence of deleted data) compared to existing methods, confirming its effectiveness. By focusing on only updating the leaf values or substituting a few isolated trees, SecureCut exhibited a significant reduction in unlearning time compared to methods that require more extensive model reconstruction or full re-training, proving its efficiency and scalability for GBDTs in VFL.

To summarize, **Sharded** excels at managing the scale of horizontal FL by balancing the computational load (via sharding) and the storage load (via coding). **KNOT**'s primary strength is mitigating the performance impact of unlearning in a high-latency, asynchronous environment by isolating the retraining to a single cluster while others continue. **SecureCut** is uniquely positioned for VFL, solving the exact unlearning problem for the highly sensitive GBDT models used in finance and healthcare, by addressing the complex, feature-based deletion requests inherent to that setting. The weaknesses are typically inherent trade-offs: Sharded and KNOT require handling the complexity of client grouping (sharding/clustering), and SecureCut is tightly coupled to the GBDT architecture, limiting its general applicability to Deep Neural Networks.

4.9 Challenges in Unlearning for Complex Model Architectures

4.9.1 The Multi-dimensional Challenge Landscape

Machine unlearning across diverse model architectures confronts a constellation of interconnected challenges spanning fundamental computational limitations, model-specific architectural constraints, verification complexities, and practical scalability requirements. Unlike many machine learning research problems exhibiting relatively straightforward optimization landscapes, unlearning presents inherent tensions between competing objectives that resist simple technical solutions. Understanding these challenges requires examining not only individual obstacles but their complex interdependencies and the irreducible trade-offs they create.

4.9.2 Fundamental Computational and Stochastic Challenges

At the foundation of unlearning difficulties lies the stochastic nature of modern machine learning training. Standard supervised learning employs stochastic gradient descent, where training data shuffles randomly across batches, mini-batch order varies across epochs, and parallelization introduces non-determinism. This stochasticity means no two training runs produce identical models despite using identical data, making the theoretical objective of achieving statistical indistinguishability between unlearned models and retrained-from-scratch models fundamentally problematic.

The mathematical consequence proves profound: when practitioners attempt to design deterministic unlearning algorithms that guarantee identical behavior to retraining, stochasticity ensures this goal remains forever unreachable. Approximate unlearning methods accommodate this reality, but only through accepting weaker guarantees, creating unavoidable information gaps between claimed and actual forgetting levels.

Data interconnections represent an equally fundamental challenge emerging from how machine learning models operate. Models do not process individual training points independently; instead, they extract intricate statistical patterns and interdependencies spanning entire datasets. When unlearning targets specific points, removing their influence necessarily disrupts these learned relationships, potentially degrading model coherence on retained data. The challenge intensifies for architectures like graphs or transformers where explicit connectivity enforces that each data point's information flows through multiple neighbors, making isolation technically impossible.

4.9.3 Model-Specific Architectural Challenges

Catastrophic forgetting—a phenomenon where aggressive unlearning induces dramatic performance collapse—emerges as a critical failure mode across architectures. This phenomenon reflects the stability-plasticity dilemma: models require sufficient plasticity to forget target information but excessive plasticity destabilizes retained knowledge. When unlearning algorithms employ large parameter updates to reverse learned associations, these updates often disrupt unrelated learned features that models depend upon for retained-data predictions.

The problem intensifies through knowledge interdependency—the reality that complex models encode information redundantly across multiple parameter regions and learned representations. Attempting to remove knowledge from isolated parameter sets risks leaving residual information in interconnected regions. For deep neural networks, knowledge distributes across millions of parameters through non-linear transformations making complete removal theoretically challenging. In transformers, knowledge embeds in attention patterns, hidden activations, and embeddings simultaneously, requiring coordinated modification across architectural components.

Parameter complexity in modern architectures creates fundamental obstacles to precise unlearning design. Deep neural networks present so-called "double descent" phenomena where increasing model capacity initially improves generalization then degrades it before improving again—a property complicating theoretical analysis. Graph neural networks introduce multi-hop propagation challenges where information about deleted nodes persists through aggregation operations across multiple layers, requiring unlearning to carefully reconstruct intermediate representations.

4.9.4 Catastrophic Collapse and Failure Modes

Research reveals that approximate unlearning methods frequently exhibit catastrophic collapse, where models produce nonsensical outputs after aggressive unlearning attempts. Gradient ascent approaches maximizing loss on forget sets often cause complete model breakdown—accuracy on retain sets drops to random-guessing levels while remaining data still trains normally.

Negative Preference Optimization (NPO) addresses this through preference-based formulations, but fundamental tensions persist. When practitioners desire strong forgetting, the necessary parameter updates often destabilize general language understanding; when they preserve utility, residual knowledge sometimes remains. This represents not an implementation limitation but a fundamental mathematical reality: completely erasing specific knowledge while preserving arbitrary other knowledge requires either complete retraining or accepting incomplete solutions.

Recent research demonstrates that even state-of-the-art unlearning methods fail systematically against adversarially-designed poisoned data. Methods that successfully remove clean training data prove ineffective at removing poisoned samples, revealing that unlearning mechanisms and data properties interact in unexpected ways. This finding suggests the field lacks complete understanding of unlearning's theoretical foundations.

4.9.5 Verification and Security Challenges

Verifying that unlearning actually removes data—as opposed to merely degrading model outputs—presents arguably the most troubling challenge. Membership inference attacks enable adversaries to statistically infer whether specific training points influenced model outputs. The ironically counterintuitive finding: models post-unlearning sometimes exhibit higher susceptibility to membership inference attacks on forget-set data, potentially because unlearning leaves distinctive artifacts.

Recent work introducing Forensic Unlearning Membership Attacks (FUMA) demonstrates that even when unlearning successfully removes retrievable knowledge, detectable artifacts enable reconstruction of forget-set membership with near-perfect accuracy in many cases. This finding reveals that unlearning creates discernible "fingerprints" in model parameter space—the very traces practitioners intended to eliminate remain identifiable through statistical analysis.

Residual information leakage represents a particularly insidious challenge because information removal proves difficult to verify. Models trained with sensitive data retain subtle statistical traces even after unlearning—models may not retrieve exact information when queried but their outputs differ measurably from models trained without that data in ways adversaries can exploit. Traditional privacy measures like differential privacy focus on bounded information leakage, but machine unlearning requires complete information removal for specific samples, a substantially different objective.

4.9.6 Computational and Practical Scalability Constraints

The fundamental computational trade-off in machine unlearning reflects the inescapable tension between exactness and efficiency. Exact unlearning through complete retraining scales linearly with forgotten-data volume, becoming impractical for large deletions. Approximate methods reduce computational costs exponentially but sacrifice unlearning guarantees.

For massive models like large language models containing billions of parameters, even approximate unlearning consumes substantial resources. Federated learning settings present additional complexity: distributed clients prevent centralized monitoring, and global model aggregation mixes influences across all participants, making client-specific unlearning substantially more difficult than centralized scenarios.

Real-time processing constraints add urgency to scalability concerns. Many applications demand near-instant unlearning responses to privacy requests, yet even approximate unlearning methods require minutes to hours for large models. This temporal tension—between computational necessity and business/regulatory demands—drives practitioners toward faster approximate methods despite inferior theoretical guarantees.

Storage complexity emerges as an often-overlooked challenge. Partition-based methods like SISA require maintaining multiple model replicas and careful bookkeeping of data-to-shard mappings, imposing substantial storage burdens for deployment. Influence function approaches cache gradient information for efficiency, further inflating memory requirements.

4.9.7 Cross-Cutting Trade-Offs Without Silver Bullets

Machine unlearning presents three fundamentally irreducible trade-offs lacking universal optimal solutions. Completeness versus efficiency forces choices between exact removal (expensive retraining) and efficient approximation (incomplete removal). Different application domains require different points on this spectrum: healthcare applications might prioritize completeness despite computational costs, while commercial recommendation systems might prioritize efficiency.

Utility preservation versus aggressiveness reflects the reality that removing information often damages model performance on retained data. Stronger unlearning causes more utility degradation; weaker unlearning preserves performance but leaves residual information. Recent stability-plasticity research suggests no unified solution exists; mechanisms preserving old knowledge often hinder new information acquisition.

Scalability versus model quality emerges in partition-based approaches: smaller partitions reduce per-request computational costs but degrade individual model quality through reduced training data and lost synergistic information. Larger partitions improve model quality but concentrate unlearning costs.

4.9.8 Unresolved Theoretical Questions

Despite substantial recent research, fundamental theoretical gaps persist. Learnability of unlearning remains unclear—whether certain data points prove inherently difficult to unlearn due to strong model dependencies. Recent work on "unlearning difficulty" via Kernelized Stein Discrepancy suggests variability, but predictive theory remains limited.

The reversibility of unlearning raises philosophical concerns: whether unlearned knowledge can be reconstructed or remains permanently inaccessible. Preliminary evidence suggests some unlearning is reversible through creative prompting, implying knowledge persists in latent form—suggesting current unlearning methods "hide" rather than "remove" knowledge.

4.9.9 Open Challenges for Practitioners

Organizations implementing unlearning face practical challenges lacking current solutions. Audit and compliance complexity emerges when demonstrating unlearning effectiveness to regulators unfamiliar with technical nuances. Membership inference attacks provide statistical evidence but not deterministic proof—regulators accustomed to exact requirements struggle with probabilistic claims. Fairness in unlearning remains underexplored: when unlearning corrupts model performance, fairness degradation often concentrates on minority groups, potentially violating fair AI principles while protecting privacy.

Challenges in machine unlearning across model architectures transcend typical machine learning obstacles through fundamental tensions between competing objectives that admit no universal solutions. Stochasticity, data interdependencies, catastrophic collapse risks, verification impossibility, and computational constraints collectively create a problem landscape requiring not breakthrough algorithms but rather nuanced engineering choices balanced against application-specific requirements. As unlearning matures from research novelty to industrial practice, successfully navigating these trade-offs will prove as important as technical algorithm design.

Chapter 5
Verification and Evaluation of Machine Unlearning Techniques

Abstract Machine unlearning has emerged as a critical mechanism for enforcing data privacy rights and removing undesired knowledge from trained machine learning models. As organizations increasingly deploy unlearning as a service to comply with regulations such as GDPR and CCPA, rigorous verification and evaluation frameworks become paramount. This chapter provides a comprehensive analysis of the multifaceted landscape of verification and evaluation metrics for machine unlearning. We systematically examine metrics for measuring unlearning effectiveness, explore statistical approaches to verification, discuss adversarial testing methodologies, analyze critical tradeoffs between model utility and unlearning guarantees, and review standardized benchmarks for evaluation. Through comparative analysis across different machine unlearning algorithms, we demonstrate how appropriate metric selection depends on algorithm characteristics, computational constraints, and application domains. This chapter synthesizes state-of-the-art research and provides practitioners with actionable guidance for designing robust unlearning systems.

Keywords Evaluation · Verification · Correctness of unlearning algorithms · Trust

5.1 Background

The exponential growth of machine learning applications has created an unprecedented data landscape. Models trained on vast datasets often inadvertently memorize sensitive personal information, proprietary data, and potentially harmful knowledge. While this memorization capability was initially viewed as beneficial for model generalization, it now poses significant privacy and security risks. The regulatory environment has fundamentally shifted with the introduction of data protection laws, most notably the European Union's General Data Protection Regulation (GDPR) and similar frameworks worldwide. These regulations grant individuals the "right to be forgotten"—a fundamental entitlement to request deletion of their personal data from both databases and the models trained upon them (Li et al. 2024; Yao et al. 2023).

R. Gupta et al., *Machine Unlearning: Concepts, Techniques and Applications*,
Studies in Computational Intelligence 1270,
https://doi.org/10.1007/978-3-032-18447-4_5

Machine unlearning addresses this requirement by enabling models to selectively forget specific data points and their associated influences without necessitating expensive retraining from scratch. However, claims of unlearning cannot be accepted at face value. Model providers may have financial or operational incentives to claim successful unlearning while preserving data influence for performance optimization. Data contributors require robust mechanisms to verify genuine deletion. Simultaneously, model developers must objectively assess whether their unlearning algorithms achieve their stated objectives while maintaining model utility on remaining data (Li et al. 2024; Zhou et al. 2024).

The evaluation and verification of machine unlearning presents unprecedented technical challenges. Unlike conventional machine learning evaluation where performance is typically measured through prediction accuracy on test data, unlearning evaluation requires fundamentally different metrics. The central question becomes: How do we verify that information has been genuinely removed from a black-box model when adversaries have incentives to report false unlearning? This chapter addresses this multifaceted challenge through comprehensive examination of verification metrics, evaluation methodologies, statistical validation approaches, adversarial robustness testing, and practical benchmarking frameworks.

5.2 Foundational Concepts for Evaluations

Before examining verification and evaluation metrics in detail, we establish fundamental definitions and properties that effective machine unlearning must satisfy. Understanding these foundational concepts provides crucial context for subsequent metric design and selection (Li et al. 2024).

5.2.1 Core Properties of Machine Unlearning

An effective machine unlearning algorithm must satisfy four essential properties:

Effectiveness: The unlearned model must completely remove all information associated with the forgotten data, rendering it indistinguishable from a model that never encountered this data. Complete information elimination ensures no residual traces of the forgotten data remain in model parameters, activations, or decision boundaries.

Efficiency: Unlearning must respond promptly to user requests within legally mandated timeframes (typically 30 days under GDPR). Computational efficiency becomes critical as unlearning requests accumulate over time in production systems. The computational advantage over naive retraining represents a primary motivation for approximate unlearning methods.

Utility Preservation: Post-unlearning model performance on remaining data should remain consistent with pre-unlearning capabilities. Unlearning processes should not degrade model accuracy or other performance metrics for tasks unrelated to the forgotten data. This property ensures that privacy objectives do not compromise the practical utility of deployed models.

Compatibility: Unlearning algorithms should be deployable across diverse model architectures and learning paradigms without requiring substantial modifications. Practical unlearning solutions must accommodate different neural network types (CNNs, RNNs, Transformers, GNNs) and machine learning frameworks (Fig. 5.1).

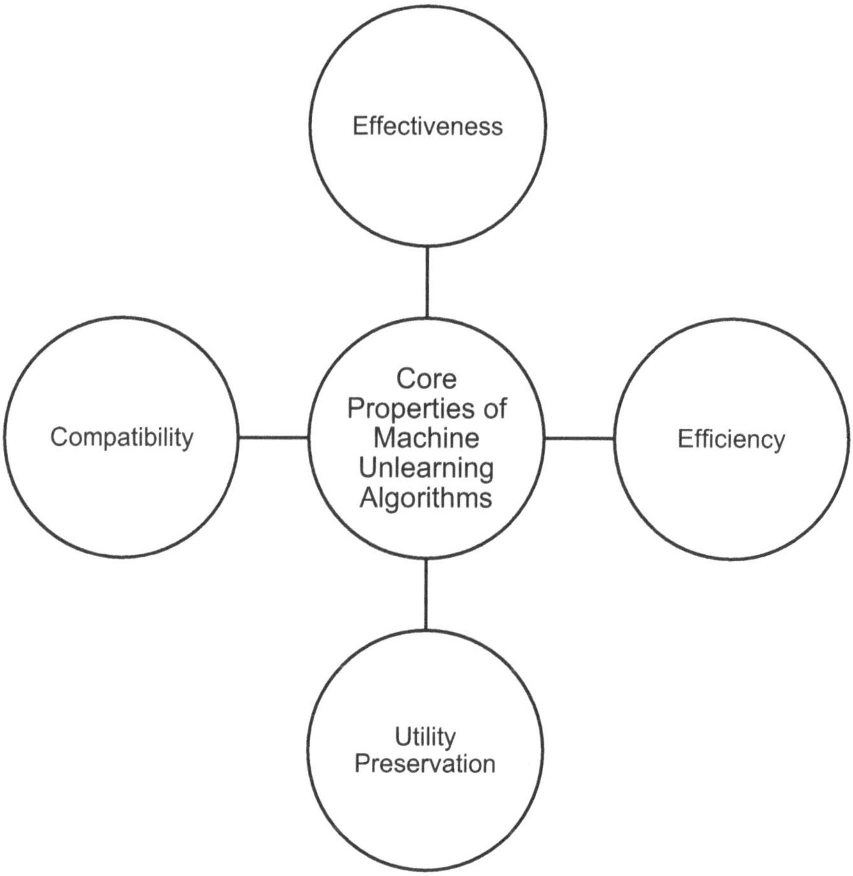

Fig. 5.1 Four core properties of any Machine Unlearning Algorithms

5.2.2 *Exact Versus Approximate Unlearning*

The machine unlearning landscape divides into two distinct paradigms with fundamentally different guarantees (Li et al. 2024). Exact unlearning involves rapid retraining on the remaining dataset after removing forgotten data, ensuring the distribution of the unlearned model is indistinguishable from a model trained from scratch without the forgotten data. Exact unlearning provides theoretically guaranteed data removal but incurs substantial computational costs, particularly for large-scale deep learning models. The SISA (Sharded, Isolated, Sliced, Aggregated) framework exemplifies exact unlearning by partitioning training data into disjoint shards and training sub-models independently (Bourtoule et al. 2021). Upon receiving an unlearning request, only the sub-model trained on the relevant shard requires retraining, substantially reducing computational burden compared to full model retraining.

Approximate unlearning avoids retraining by directly modifying model parameters to achieve statistical indistinguishability from the retrained model. This approach substantially reduces computational costs but provides weaker privacy guarantees. Approximate unlearning methods employ diverse strategies: gradient ascent that maximizes loss on forgotten data, influence function-based approaches that estimate and negate data influence, and knowledge distillation methods that transfer non-forgotten knowledge to new models while filtering sensitive information.

5.3 Verification Metrics: Ensuring Genuine Data Removal

Verification metrics enable data contributors to assess whether models have genuinely forgotten their data. These metrics address a critical asymmetry: model providers have incentives to report unlearning success while conserving computational resources, but data contributors require mechanisms to verify authentic data removal (Li et al. 2024). Verification metrics divide into two categories based on their impact on model development (Fig. 5.2).

5.3.1 *Invasive Verification Metrics*

Invasive metrics require data contributors to actively watermark data before model training—a process that influences model development. These metrics operate through two distinct phases.

Watermarking Phase: Data contributors embed special markers into data destined for potential unlearning. Common watermarking approaches include backdoor-based watermarks that modify prediction patterns for marked data points, feature injection

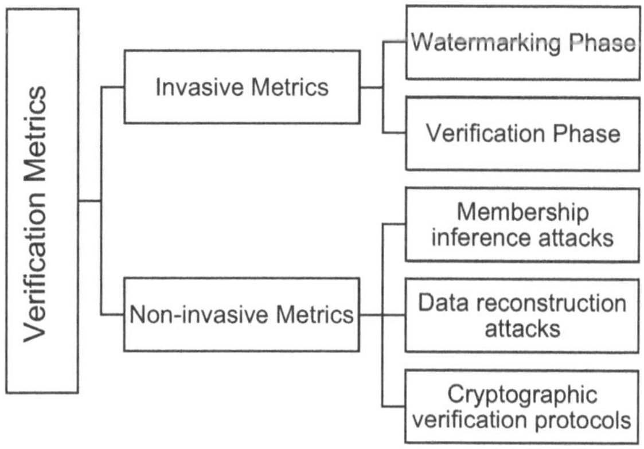

Fig. 5.2 Properties and phases for verification metrics

that adds synthetic features with deliberate patterns, and adversarial example-based watermarks that characterize decision boundaries through specially crafted inputs.

Verification Phase: After claimed unlearning completion, data contributors analyze whether watermarks persist in the unlearned model. Backdoor-based verification measures the success rate of poisoned inputs in triggering artificial target labels. Feature-based verification checks whether injected features have been eliminated from model weights. Adversarial watermark verification assesses whether the model still manifests smoothed decision boundaries around watermarked adversarial examples.

The primary advantage of invasive metrics lies in intuitive interpretation and relatively straightforward implementation. However, invasive approaches impose significant costs. Watermarking intrudes into the model development process and may degrade model performance. More critically, adversarial watermarks that model providers become aware of can be deliberately circumvented through defensive mechanisms (Guo et al. 2024). Backdoor-based watermarks remain particularly vulnerable to detection and exploitation by malicious actors.

5.3.2 Non-invasive Verification Metrics

Non-invasive metrics operate without requiring advance watermarking, enabling verification for legacy models where data contributors did not previously insert watermarks. These metrics extract verification information from model outputs or parameters.

Membership Inference Attacks: Membership inference determines whether specific data samples existed in a model's training set. This approach leverages the observation that models typically exhibit higher confidence on training data than non-training data. A binary meta-classifier trained on model outputs can distinguish between members and non-members. If the unlearned model exhibits substantially reduced membership inference attack success rates for forgotten data, effective unlearning is indicated. The Forgetting Rate metric quantifies the change in membership inference precision from before to after unlearning, with a Forgetting Rate of 1.0 indicating complete transition of forgotten data from member to non-member classification (Liu et al. 2024).

Data Reconstruction Attacks: Data reconstruction attempts to reverse-engineer training data from model parameters or outputs. For class-level unlearning, gradient descent optimization can reconstruct class-representative images from model gradients. If reconstructed data from unlearned models no longer contains information about forgotten classes, successful unlearning is indicated. However, reconstruction attacks remain computationally expensive and less well-understood for large language models compared to vision models.

Cryptographic Verification Protocols: Eisenhofer and colleagues developed the first verifiable unlearning protocol employing succinct non-interactive zero-knowledge arguments. This approach enables model providers to offer cryptographic proof that prescribed unlearning was executed. Cryptographic verification provides theoretical guarantees superior to statistical approaches but faces challenges in practical deployment due to computational overhead and the requirement for sophisticated cryptographic infrastructure (Eisenhofer et al. 2023).

5.4 Evaluation Metrics: Assessing Unlearning Quality

While verification metrics answer whether unlearning occurred, evaluation metrics assess the quality of unlearning implementation from the model provider's perspective. Evaluation metrics divide into three categories corresponding to the three performance dimensions of unlearning algorithms: effectiveness, efficiency, and utility preservation (Li et al. 2024; Yao et al. 2023).

5.4.1 Effectiveness Metrics

Effectiveness metrics quantify the degree to which unlearned models lack information about forgotten data.

Relearn Time Analysis: The relearn time metric measures the number of training epochs required for an unlearned model to regain performance equivalent to the original model on forgotten data. Short relearn times indicate substantial information retention within the unlearned model. Extended relearn times suggest genuine knowledge elimination. The Anamnesis Index (AIN) refines this approach by introducing a margin around original accuracy levels. AIN approaching 1.0 indicates effective unlearning, while values substantially exceeding 1.0 suggest over-unlearning with potential performance degradation (Chundawat et al. 2023) (Fig. 5.3).

Similarity-Based Metrics: The similarity between unlearned and retrained models provides a proxy for unlearning effectiveness. This comparison operates at three levels with increasing discriminative power (Li et al. 2024):

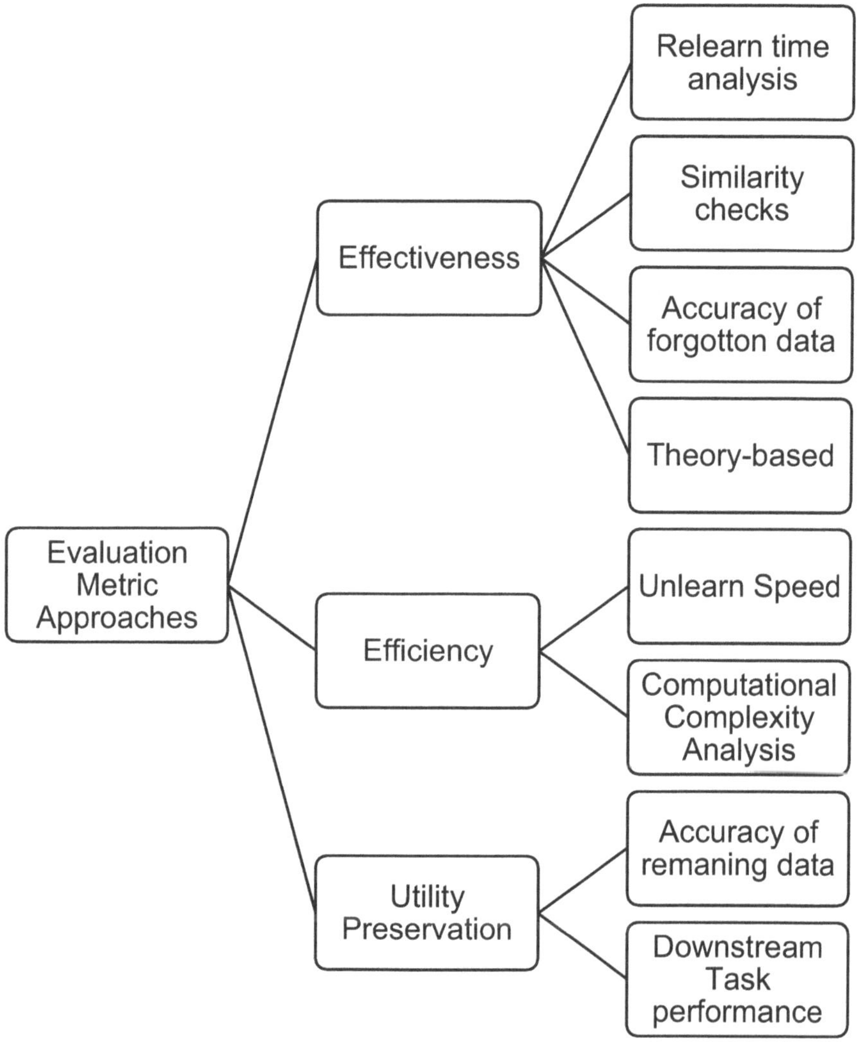

Fig. 5.3 Areas of consideration for evaluation metric approaches

- Activation Distance: L1 and L2 norm distances measure differences in final layer activations between unlearned and retrained models. Lower activation distances indicate closer correspondence.
- Weight Distance: L2-norm distance and cosine similarity quantify weight parameter differences. Cosine similarity correlates with unlearning effectiveness but applies only to classification tasks.
- Distribution Distance: Kullback–Leibler (KL) divergence and Jensen-Shannon divergence measure output distribution differences. These metrics provide stronger unlearning guarantees than activation or weight distances, though all similarity-based metrics require access to the retrained model, which may be computationally prohibitive.

Accuracy on Forgotten Data: Simple measurement of model accuracy on forgotten data provides the most straightforward effectiveness proxy. Models demonstrating near-zero accuracy on forgotten samples while retaining high accuracy on remaining data have successfully unlearned. However, accuracy-based metrics lack sensitivity to subtle information retention mechanisms and may conflate poor generalization with successful unlearning.

Theory-Based Metrics: Certain unlearning algorithms inherently prove their effectiveness through mathematical construction. Exact unlearning methods guarantee effectiveness through retraining procedures. Certified unlearning approaches add controlled noise to model weights or loss functions based on differential privacy theory, offering mathematical proofs that unlearned models are indistinguishable from retrained models. These theoretical guarantees provide the strongest effectiveness assurance but apply only to specific algorithm classes and often fail for non-convex deep learning models.

5.4.2 *Efficiency Metrics*

Efficiency metrics quantify computational resource consumption during unlearning operations. With unlearning requests accumulating over the model lifecycle, efficiency becomes critical.

Unlearn Speed: The runtime comparison between unlearning and naive retraining provides the fundamental efficiency metric. Speedups exceeding 100–1000 $\times$ over retraining represent the principal motivation for approximate unlearning research. However, direct runtime comparisons face challenges: naive retraining may be computationally infeasible for large models, and runtime varies significantly across hardware platforms.

Computational Complexity Analysis: Theoretical analysis of algorithmic complexity provides hardware-independent efficiency assessment. Methods based on influence functions or Hessian matrix operations often demonstrate quadratic computational growth with dataset size, imposing scalability constraints. Gradient-based approaches typically exhibit linear complexity, enabling better scalability but potentially sacrificing effectiveness.

5.4.3 *Utility Preservation Metrics*

Utility preservation ensures that unlearning objectives do not compromise model functionality on remaining data and unrelated tasks.

Accuracy on Remaining Data: The most direct utility metric measures classification or task performance on data outside the unlearning scope. Model accuracy on test sets derived from the remaining training dataset provides the primary utility indicator. High accuracy on remaining data combined with low accuracy on forgotten data signals successful unlearning with preserved utility.

Downstream Task Performance: For language models and multimodal systems, utility extends beyond primary task accuracy to broader capabilities. Benchmarks measuring natural language understanding, reasoning ability, instruction-following performance, and common-sense reasoning provide comprehensive utility assessment. Unlearning algorithms causing substantial utility degradation on these downstream tasks exhibit undesirable tradeoffs (Liu et al. 2024; Yao et al. 2023).

5.5 Statistical Approaches to Unlearning Verification

Statistical hypothesis testing provides rigorous frameworks for unlearning verification, moving beyond simple metric thresholds to formal probabilistic inference about data removal success.

5.5.1 *Membership Inference as Statistical Testing*

Membership inference attack success constitutes a statistical test for data retention. The fundamental hypothesis becomes:

- Null Hypothesis (H_0): The unlearned model exhibits no greater membership inference attack success on forgotten data than random guessing
- Alternative Hypothesis (H_1): The unlearned model demonstrates statistically significant membership inference attack success on forgotten data

Performing membership inference attacks before and after unlearning enables computation of test statistics and confidence intervals around attack accuracy. Substantial accuracy reductions following unlearning provide evidence for H_0.

However, computational expense limits membership inference applicability—training shadow models and meta-classifiers requires substantial resources.

5.5.2 *Statistical Hypothesis Testing on Accuracy Differences*

The difference in model accuracy on forgotten vs. remaining data provides another statistical testing framework. For classification tasks:

Let p_f represent classification accuracy on forgotten data and p_r represent accuracy on remaining data. The hypothesis test becomes:

- Null Hypothesis: p_f = random_guess_probability (model has unlearned forgotten data)
- Alternative Hypothesis: p_f > random_guess_probability (model retains information about forgotten data)

Binomial hypothesis testing or more sophisticated statistical frameworks can determine whether observed accuracy differences exceed those expected from random variation. Confidence intervals around accuracy estimates quantify uncertainty in unlearning assessment.

5.5.3 *Confidence and Credibility in Unlearning Claims*

Because approximate unlearning provides only statistical indistinguishability rather than perfect unlearning, quantifying confidence in unlearning claims becomes essential. Bayesian frameworks enable integration of prior beliefs with observed data to produce posterior probabilities that unlearning succeeded (Gao et al. 2024).

5.6 Adversarial Testing and Robustness Evaluation

While standard metrics assess unlearning under controlled conditions, adversarial testing explores whether unlearning persists against determined attacks. This critical dimension often reveals that naive unlearning approaches fail when subjected to sophisticated extraction attacks.

5.6.1 Membership Inference Attacks

Membership inference attacks constitute the primary adversarial testing methodology. Modern attacks employ diverse strategies beyond simple confidence-based thresholds (Duan et al. 2024):

- Likelihood-based approaches that analyze model probability distributions
- Loss-based methods that exploit differences in cross-entropy loss between training and non-training data
- Gradient-based techniques that fingerprint model responses to specific inputs
- Language model-specific attacks that leverage textual generation patterns

Comprehensive adversarial evaluation applies multiple membership inference variants to ensure unlearning robustness across attack methodologies.

5.6.2 Model Inversion Attacks

Model inversion attacks reconstruct training data from model parameters or predictions. For classification tasks, gradient-based optimization can synthesize inputs that maximize activation for target classes. If the unlearned model permits reconstruction of forgotten class-representative data, information retention is indicated (Salem et al. 2021).

5.6.3 Relearning Attacks

Relearning attacks assess whether forgotten knowledge can be recovered through additional training on adversarial data. Recent research demonstrates that unlearned models often relearn forgotten information through few-shot fine-tuning or small-scale retraining. If relearning on forgotten data recovers model performance substantially faster than training from scratch, information retention persists despite unlearning claims (Lynch et al. 2024; Hu et al. 2024).

5.6.4 Jailbreaking and Extraction Attacks

For language models, adversarial prompting and sophisticated extraction techniques may elicit forgotten information that unlearning algorithms failed to suppress. Testing unlearned models against jailbreaking attacks and prompt injection methodologies provides practical evaluation of unlearning robustness in realistic adversarial scenarios.

5.7 Understanding the Utility-Unlearning Tradeoff

Perfect unlearning that completely eliminates all traces of forgotten data while maintaining perfect model utility on remaining data represents an unattainable ideal. In practice, all unlearning approaches involve fundamental tradeoffs between unlearning completeness and utility preservation.

5.7.1 Causes of Utility Degradation

Utility degradation during unlearning emerges from multiple sources:

Parameter Overfitting: Approximate unlearning methods directly modify model parameters to negate forgotten data influence. These parameter modifications may impair generalizable patterns learned from remaining data, particularly when forgotten and remaining data exhibit statistical dependence or feature overlap.

Catastrophic Forgetting: Aggressive unlearning modifications can catastrophically collapse model performance on both forgotten and remaining data. This phenomenon occurs when parameter updates affect broadly-used features that support multiple tasks and downstream capabilities.

Knowledge Entanglement: Deep learning models develop entangled representations where knowledge about forgotten data becomes inseparable from knowledge about remaining data. Unlearning approaches cannot cleanly extract forgotten information without collateral damage to remaining knowledge (Maini et al. 2024).

5.7.2 Quantifying Tradeoffs Through Pareto Frontiers

Rather than optimizing single metrics, comprehensive unlearning evaluation should characterize Pareto optimal tradeoff frontiers between unlearning effectiveness and utility preservation. This approach acknowledges that different applications prioritize these objectives differently. Privacy-critical applications prioritize aggressive unlearning despite utility costs, while high-performance systems tolerate less complete unlearning to maintain utility.

5.7.3 Localized Versus Distributed Unlearning Effects

Recent research reveals that unlearning effects distribute non-uniformly across model layers and components. Localized unlearning operations confined to few parameters or layers may achieve effective forgotten data removal while preserving utility.

Conversely, distributed modifications across many layers often cause collateral utility degradation. Understanding and exploiting this localization property enables improved tradeoffs (Gao et al. 2024).

5.8 Standardized Benchmarks for Unlearning Evaluation

Standardized benchmarks enable systematic comparison of unlearning algorithms and reproducible evaluation across research groups. Several influential benchmarks have emerged.

5.8.1 TOFU Benchmark

The TOFU (Task Of Fictitious Unlearning) benchmark addresses a critical evaluation challenge: unlearning large language models against retraining remains computationally prohibitive. TOFU creates synthetic scenarios by injecting fictitious author biographies into model training data. Since these biographies never appeared in the original pretraining corpus, fine-tuning on the fictional dataset simulates retraining. This enables practical comparison of unlearned model performance against retrained baselines (Maini et al. 2024).

The benchmark evaluates unlearning effectiveness through question-answering tasks about fictional authors. Effective unlearning exhibits low accuracy on forgotten author questions. Utility preservation is assessed through general knowledge and reasoning benchmarks.

5.8.2 WMDP Benchmark

The Weapons of Mass Destruction Prevention (WMDP) benchmark evaluates unlearning of hazardous knowledge spanning cybersecurity, chemistry, and biology domains. This benchmark directly addresses safety–critical unlearning scenarios where models must forget dangerous information. Assessment involves zero-shot question answering about hazardous content. Effective unlearning demonstrates substantially reduced accuracy on harmful questions while maintaining capabilities on benign topics (Li et al. 2024).

5.8.3 MUSE Benchmark

The MUSE (Machine Unlearning Six-Way Evaluation) benchmark provides comprehensive evaluation across six distinct unlearning scenarios encompassing privacy, copyright, and safety concerns. This multi-dimensional benchmark reveals that single unlearning approaches often perform inconsistently across diverse scenarios. The benchmark includes news articles, books, and factual information as unlearning targets.

5.9 Comparative Analysis: Metric Suitability Across Algorithms

Different unlearning algorithms exhibit varying characteristics that determine appropriate evaluation metrics.

5.9.1 Metrics for Exact Unlearning Algorithms

Exact unlearning methods including SISA and retraining-based approaches inherently guarantee effectiveness through mathematical construction. For these methods:

- Theory-based metrics directly apply, providing formal effectiveness guarantees
- Similarity metrics comparing to retrained models become highly relevant
- Efficiency metrics should emphasize speedup relative to full retraining
- Accuracy on forgotten data becomes almost meaningless—exact unlearning mathematically guarantees near-zero accuracy

Exact unlearning's primary evaluation focus should concentrate on computational efficiency gains and utility preservation rather than effectiveness verification.

5.9.2 Metrics for Approximate Gradient-Based Methods

Gradient-based approximate unlearning (gradient ascent, gradient descent on relabeled data) operates by modifying loss landscapes. For these methods:

- Effectiveness metrics become critical—gradient-based approaches provide no theoretical guarantees
- Relearn time analysis provides strong indicators of gradient-based unlearning success
- Adversarial testing, particularly relearning attacks, reveals vulnerabilities in gradient-based approaches

- Membership inference attacks often succeed against gradient-based unlearning

Gradient-based methods exhibit particular vulnerability to relearning attacks, suggesting comprehensive adversarial evaluation as essential.

5.9.3 Metrics for Influence Function-Based Methods

Influence function approaches estimate and negate the impact of forgotten data through second-order derivative information. These methods benefit from:

- Distribution distance metrics that directly measure influence elimination
- Theory-based metrics grounded in influence function mathematics
- Parameter sensitivity analysis quantifying how removing data influence affects specific weights

Computational complexity of Hessian approximation necessitates emphasis on efficiency metrics and scalability analysis.

5.9.4 Metrics for Knowledge Distillation Methods

Knowledge distillation-based unlearning transfers non-forgotten knowledge to new models while filtering sensitive information. These approaches require:

- Utility preservation metrics become paramount—knowledge distillation inherently risks information loss
- Evaluation on held-out data from the remain set to assess transfer quality
- Downstream task performance across diverse benchmarks

Table 5.1 summarizes metric suitability across different unlearning algorithm categories, enabling practitioners to select appropriate evaluation strategies.

5.10 Challenges and Implications

5.10.1 Future Challenges

Despite substantial progress in unlearning metrics and evaluation, significant challenges persist.

Evaluation Under Distribution Shift—Current evaluation protocols primarily assess unlearning on test data similar to training distribution. Practical evaluation

Table 5.1 Summary of metric suitability for different unlearning algorithms

Metric	Effectiveness	Scalability	Interpretability	Computational cost	Applicable algorithms
Relearn time	High	Medium	High	High	Approximate methods
Accuracy on D_f	High	High	Very high	Very low	All
Similarity-based	Very high	Medium	Medium	Medium	Approximate, Exact
Membership inference	Medium	Low	Medium	Very high	All
Data reconstruction	High	Low	High	Very high	Vision models
Theory-based	Very high	High	Low	Low	Certified methods
Unlearn speed	N/A	High	Very high	Low	All
Downstream task	Medium	High	Medium	Medium	LLM unlearning
PCA analysis	High	Medium	Very high	Medium	Approximate
CKA analysis	High	Medium	Very high	Medium	Approximate
Fisher information	Very high	Low	Medium	High	Approximate

should characterize unlearning robustness under distribution shift—do unlearning guarantees persist when deployment conditions differ from evaluation scenarios?

Continual and Sequential Unlearning—Real-world systems receive unlearning requests continuously. Evaluation frameworks should assess how unlearning performance degrades as request counts accumulate. Current metrics typically evaluate single unlearning requests rather than sequences.

Multimodal and Cross-Modal Unlearning—As models incorporate multiple modalities (vision-language models, audio-visual systems), evaluation must characterize unlearning effects across modalities. Does unlearning information in one modality implicitly affect related information in other modalities?

Formal Verification and Provable Unlearning—Advancing from probabilistic verification to formal proofs of unlearning represents a critical frontier. Integration of formal verification methods with machine learning could enable mathematically rigorous unlearning guarantees applicable to practical deep learning systems.

5.10.2 *Implications*

Model providers implementing unlearning should:

1. Implement multi-level evaluation combining effectiveness, efficiency, and utility metrics
2. Conduct adversarial evaluation including membership inference and relearning attacks
3. Maintain comprehensive documentation of evaluation methodologies and metric values
4. Regularly validate unlearning through ongoing verification procedures
5. Establish fallback procedures when unlearning evaluation fails
6. Provide transparency to data subjects regarding verification methodologies employed.

Data contributors and regulators should:

1. Demand evidence of comprehensive evaluation rather than accepting claims alone
2. Require disclosure of specific metrics, thresholds, and evaluation protocols
3. Prioritize independent verification mechanisms rather than relying solely on provider claims
4. Establish minimum standards for unlearning evaluation across regulatory jurisdictions
5. Invest in public verification infrastructure and open-source evaluation tools
6. Maintain skepticism toward metrics showing suspiciously perfect unlearning performance.

The research community should prioritize:

1. Developing formal verification frameworks bridging the gap between theoretical guarantees and practical systems
2. Creating unified benchmarking standards enabling fair algorithm comparison
3. Investigating mechanistic interpretability approaches to understand unlearning mechanisms
4. Designing efficient adversarial testing methodologies scalable to billion-parameter models
5. Studying continual unlearning as opposed to single-request scenarios
6. Exploring cross-modal and multimodal unlearning evaluation.

5.10.3 Conclusion

Verification and evaluation of machine unlearning represents a critical and multi-faceted challenge at the intersection of privacy, machine learning, and cryptography. This chapter has systematically examined the landscape of metrics, methodologies, and frameworks for assessing unlearning quality.

Effective unlearning verification requires multiple complementary approaches rather than reliance on single metrics. Invasive verification mechanisms provide strong guarantees when applicable, while non-invasive approaches enable verification for legacy systems. Evaluation metrics addressing effectiveness, efficiency, and utility enable comprehensive algorithm assessment. Adversarial testing reveals vulnerabilities in naive approaches, while mechanistic interpretability tools distinguish genuine information removal from superficial metric manipulation.

The fundamental tension between unlearning completeness and utility preservation necessitates contextual metric selection aligned with application priorities. High-stakes privacy applications demand aggressive verification despite utility costs, while high-performance systems tolerate less complete unlearning. Standardized benchmarks including TOFU, WMDP, and MUSE enable reproducible evaluation and systematic algorithm comparison.

Moving forward, the field must advance from single-metric evaluation toward comprehensive verification ecosystems incorporating formal verification, adversarial robustness assessment, and mechanistic understanding of unlearning mechanisms. As regulatory requirements intensify and machine learning systems grow increasingly capable and consequential, rigorous unlearning verification transforms from research interest to operational necessity.

The path toward trustworthy machine unlearning lies not in perfect metrics—which may not exist—but in transparent, comprehensive, and adversarially robust evaluation frameworks that provide genuine assurance to all stakeholders that forgotten data remains forgotten.

Chapter 6
Applications of Machine Unlearning

Abstract This chapter offers a comprehensive analysis of machine unlearning—the targeted removal of specific learned knowledge from AI models—focusing on practical applications across enterprise settings. It reviews leading platforms including Hirundo, IBM, Microsoft, Google, Cleanlab, and AWS, and details their specialized unlearning solutions that address regulatory compliance, privacy preservation, model remediation, and operational efficiency. Through in-depth case studies of global (JPMorgan Chase, Amazon, Microsoft Azure) and Indian (State Bank of India, Paytm, Reliance Jio) enterprises, the chapter demonstrates measurable business impact such as cost reduction, improved compliance response times, enhanced trust, and sustained model accuracy. It highlights deployment in diverse business functions—HR, finance, IT, operations, and customer service—illustrating how unlearning enables privacy-respecting talent management, regulatory-aligned financial analytics, resilient cybersecurity, adaptive manufacturing, and customer data minimization without retraining models from scratch. The chapter emphasizes the importance of "privacy-by-design" architectures, the critical balance among fidelity, certifiability, and efficiency, and the necessity of robust verification protocols. It concludes with key learnings on overcoming data dependency, integration of legacy systems, and establishing organizational policies for scalable and compliant machine unlearning deployments in real-world enterprise environments.

Keywords Machine unlearning · Enterprise AI · Regulatory compliance · Privacy preservation · Model remediation

6.1 Enterprise Platforms and Products

Machine unlearning, the process of making AI models selectively "forget" specific data or behaviors without complete retraining, has emerged as a critical capability for enterprise AI deployments. While **Hirundo** stands as the first dedicated machine

R. Gupta et al., *Machine Unlearning: Concepts, Techniques and Applications*,
Studies in Computational Intelligence 1270,
https://doi.org/10.1007/978-3-032-18447-4_6

Table 6.1 Enterprise platforms and products around machine unlearning

Product/provider	Key machine unlearning capabilities	Enterprise focus
Hirundo	Dedicated platform for comprehensive AI model unlearning	Enterprise AI remediation
IBM	Sensitive data removal, exact and approximate unlearning	AI/LLM compliance, privacy
Microsoft	Selective forgetting, bias/data removal for Azure ML	Cloud, enterprise AI, bias
Google	Unlearning algorithms, privacy/copyright issue remediation	LLMs, cloud AI, compliance
CleanLab	Data preparation for machine unlearning	Data maturity
Pathway	Data processing for machine unlearning	Realtime data processing
VE3	Custom unlearning suite (gradient/layer/data partitioning)	Data privacy, regulation
BotStocks	Memory management for Machine unlearning	Context building for AI applications
Amazon web services (AWS)	Platform services for running machine unlearning algorithms	Compute and infra
Pecan AI	Predictive analytics, data removal requests	Analytics, regulated sectors
EUPG (URV)	Privacy-preserving efficient unlearning framework	AI/ML research, industry

unlearning platform, several other companies and solutions are addressing the implementation of machine unlearning for enterprises. IBM Watson, Microsoft Research and Google Enterprise AI Services are working on various Enterprise AI Platform Solutions with Unlearning capabilities. Cleanlab and Pathway are also working on developing Data-centric AI and quality platforms. VE3 and BotStacks are also AI Platform providers with Unlearning support, AWS is a ML services based cloud provider having Unlearning support (Table 6.1).

6.1.1 Hirundo

Hirundo is the pioneering commercial machine unlearning platform, offering comprehensive solutions for enterprise AI model remediation. Founded in 2023 by former Technion Dean of Computer Science Prof. Oded Shmueli, Rhodes Scholar Ben Luria, and data lineage expert Michael Leybovich, Hirundo raised $8 million in seed funding in 2025.

Key Capabilities:

- **LLM Behavior Unlearning**: Mitigates biases (up to 70% reduction), removes hallucinations (55% reduction), and eliminates jailbreaks (85% reduction)
- **LLM Data Unlearning**: Removes PII and sensitive data with 100% removal guaranteed at scale
- **Data QA**: Detects and resolves mislabels across vision, radar, timeseries, NLP, STT, and tabular data

Hirundo's technology works across both generative and non-generative systems and is already being piloted by multinationals in finance, healthcare, and defense. The platform operates like "AI model neurosurgery," pinpointing where problematic content originates in a model's parameters and precisely removing it.

6.1.2 IBM Watson and Research Initiatives

IBM has pioneered machine unlearning techniques across its enterprise AI platforms, enabling clients to comply with data privacy regulations4. IBM's approach includes:

- **Watson Platform Integration**: Incorporates unlearning capabilities that allow clients in healthcare and finance to remove patient or customer data from models on demand
- **Granite LLM Family**: IBM is experimenting with applying unlearning to its Granite family of LLMs, using unlearning as another way to remove unwanted behavior after data filters
- **Research Leadership**: IBM researchers have published extensively on LLM unlearning methods, focusing on removing unwanted data influence without degrading model performance.

6.1.3 Microsoft Research and Enterprise Solutions

Microsoft has developed significant machine unlearning research and practical applications:

- **Harry Potter Case Study**: Microsoft tested unlearning algorithms by measuring how well they could forget copyrighted Harry Potter books while preserving related general knowledge
- **Azure Machine Learning Integration**: While not explicitly marketed as unlearning, Microsoft's Azure ML platform provides capabilities for model retraining and data management that support unlearning workflows.

6.1.4 Google Enterprise AI Services

Google has been a major contributor to machine unlearning research and development:

- **Machine Unlearning Challenge**: Google launched the first Machine Unlearning Challenge to develop techniques for removing sensitive data from AI algorithms
- **Enterprise Applications**: Google implements machine unlearning for bias mitigation in large language models, using algorithms to remove problematic data that perpetuates stereotypes.

6.1.5 Cleanlab

Cleanlab provides data-centric AI solutions that complement machine unlearning efforts. While not exclusively focused on unlearning, Cleanlab's capabilities support the broader goal of improving AI model reliability:

- **Automated Data Issue Detection**: Finds and fixes errors in ML datasets, including label errors, outliers, and duplicates
- **Cleanlab Studio**: No-code platform for finding and fixing issues in datasets, which can support unlearning preparation workflows
- **Enterprise Adoption**: Used by Fortune 500 companies to improve dataset quality and model performance.

6.1.6 Pathway

Pathway offers a real-time data processing engine with machine unlearning capabilities:

- **Real-Time Unlearning**: Uniquely enables machines to "learn to forget" through continuous data stream processing
- **Incremental Updates**: Allows revision of specific data points without requiring full batch data uploads, comparable to updating one cell in Excel without reprocessing the entire document
- **Industry Applications**: Particularly valuable for manufacturing, financial services, and logistics where real-time data accuracy is critical.

6.1.7 VE3

VE3 offers machine unlearning solutions as part of their broader AI services portfolio:

- **Bias and Toxicity Removal**: Applies machine unlearning to remove bias, toxic behaviors, and poisoned data from AI models
- **Multi-layered Approach**: Uses gradient reversal, selective weight adjustment, and layer-specific unlearning techniques
- **Enterprise Integration**: Provides machine unlearning as part of comprehensive AI ethics and responsibility solutions.

6.1.8 BotStocks

BotStacks incorporates memory management and forgetting capabilities in their AI assistant platform:

- **Adaptive Memory Systems**: Implements smart context management that can prioritize critical information and trim redundant data
- **Self-Cleaning Memory**: Automatically archives outdated information to keep AI assistants sharp
- **Enterprise Deployment**: Provides no-code AI assistant development with built-in memory optimization.

6.1.9 Amazon Web Services (AWS)

While AWS doesn't offer dedicated machine unlearning services, their comprehensive ML platform supports unlearning implementations:

- **Amazon SageMaker**: Provides the infrastructure and tools needed to implement custom unlearning solutions
- **Privacy-Focused Research**: Amazon Science has published research on privacy-adhering machine unlearning in NLP tasks
- **Scalable Infrastructure**: Offers the computational resources necessary for enterprise-scale unlearning operations.

6.2 Application Across Business Functions

6.2.1 Human Resources: Privacy-Preserving Talent Management

In Human Resources, machine unlearning addresses critical privacy challenges inherent in AI-driven recruitment, performance evaluation, and workforce analytics systems. When organizations deploy ML models for resume screening, candidate assessment, or employee churn prediction, they accumulate sensitive personal information that may require removal upon request or employment termination.

Machine unlearning enables HR departments to comply with data protection regulations by selectively removing individual employee or candidate data from trained models without compromising overall system performance. For example, when a job applicant withdraws their application or an employee exercises their right to erasure, HR systems can unlearn that individual's data while maintaining predictive accuracy for other candidates or employees. A multinational auditing company successfully implemented AI-driven CV evaluation systems with integrated unlearning capabilities, allowing automatic removal of applicant data while preserving the model's ability to assess future candidates fairly.

Beyond regulatory compliance, machine unlearning in HR facilitates bias mitigation. Organizations can identify and remove biased training samples that perpetuate discrimination based on protected characteristics such as age, gender, or ethnicity. By unlearning specific features or data points that introduce unfairness, HR systems can progressively improve their equity while avoiding the resource-intensive process of complete model retraining. This capability proves particularly valuable when organizations discover that historical hiring data reflects past discriminatory practices that should not influence future decisions.

6.2.2 Finance: Regulatory Compliance and Risk Management

The financial services sector faces stringent regulatory requirements and rapidly evolving risk landscapes, making machine unlearning particularly valuable for credit risk assessment, fraud detection, and customer analytics. Financial institutions deploy ML models for credit scoring, loan approval, and risk prediction using vast amounts of customer data. When customers close accounts, request data deletion, or when regulatory requirements mandate data removal, financial organizations must ensure complete erasure from all systems, including trained ML models.

Machine unlearning techniques enable banks and financial institutions to remove specific customer data from credit risk models without sacrificing predictive performance on the remaining customer base. For instance, a consortium of banks collaborating through federated learning to develop credit risk predictors can implement unlearning mechanisms that allow individual customers to withdraw their data contributions at contract termination while maintaining model utility for other participants.

Fraud detection systems particularly benefit from machine unlearning by enabling rapid response to data poisoning attacks. Malicious actors may attempt to inject fraudulent transaction patterns into training data to manipulate model behavior. Machine unlearning allows financial institutions to identify and remove poisoned data samples, restoring model integrity without the prohibitive costs and time delays associated with complete retraining. This capability proves essential in maintaining the security and reliability of real-time fraud prevention systems that protect billions of dollars in transactions.

Furthermore, machine unlearning facilitates compliance with evolving financial regulations. As regulatory frameworks change or institutions update their risk assessment criteria, unlearning enables selective removal of outdated data or deprecated features from models. This dynamic adaptability ensures that financial ML systems remain aligned with current regulatory standards and business requirements without constant full-scale model reconstruction.

6.2.3 Information Technology: Security and System Integrity

IT departments leverage machine unlearning primarily for cybersecurity, system optimization, and data governance. Security applications benefit significantly from unlearning's ability to mitigate the effects of adversarial attacks and data poisoning. When IT security systems detect compromised or maliciously injected data that has influenced intrusion detection models or threat assessment algorithms, unlearning techniques enable rapid remediation by removing the contaminated data's influence without disrupting ongoing security operations.

Network management and performance optimization represent another critical IT application area. Machine learning models deployed for traffic management, load balancing, and infrastructure optimization continuously adapt to changing network conditions. When data becomes outdated due to infrastructure changes, equipment upgrades, or shifts in usage patterns, machine unlearning allows IT systems to remove obsolete information while maintaining predictive accuracy for current conditions. This capability proves particularly valuable for telecommunications providers managing complex, evolving networks where complete model retraining would be operationally disruptive and prohibitively expensive.

Data governance and compliance management constitute fundamental IT responsibilities that machine unlearning directly supports. As organizations implement comprehensive data protection frameworks to comply with GDPR, CCPA, and emerging privacy legislation, IT departments must ensure that data deletion requests extend beyond databases to encompass all ML models trained on that data. Machine unlearning provides IT teams with the technical capability to implement "right to be forgotten" requests systematically across enterprise ML systems, demonstrating compliance to regulators and building stakeholder trust.

Additionally, machine unlearning enhances model maintenance efficiency in IT operations. Rather than maintaining resource-intensive model versioning systems or performing frequent complete retraining cycles, IT departments can implement targeted unlearning procedures to update models incrementally as data requirements change. This approach reduces computational overhead, minimizes energy consumption, and contributes to more sustainable AI operations—an increasingly important consideration as organizations pursue green computing initiatives.

6.2.4 Operations: Efficiency and Adaptive Manufacturing

Operations management, particularly in manufacturing and supply chain optimization, increasingly relies on ML models for predictive maintenance, quality control, demand forecasting, and process optimization. Machine unlearning enables operations teams to maintain model accuracy and relevance in dynamic production environments without the operational disruptions associated with complete model retraining.

Predictive maintenance systems exemplify machine unlearning's operational value. These systems analyze sensor data from industrial equipment to predict failures and optimize maintenance schedules, thereby reducing unplanned downtime and extending equipment lifespan. When equipment undergoes significant modifications, upgrades, or replacements, historical data from previous configurations becomes irrelevant or potentially misleading. Machine unlearning allows operations teams to remove outdated sensor data and recalibrate predictive models to reflect current equipment characteristics, ensuring maintenance predictions remain accurate and actionable.

Quality control applications similarly benefit from unlearning capabilities. Manufacturing ML models trained to detect product defects or process anomalies must adapt when production processes change, materials are substituted, or quality standards are updated. Rather than collecting entirely new training datasets and retraining from scratch—a time-consuming process that delays production optimization—manufacturers can use machine unlearning to selectively remove data related to obsolete processes or materials while retaining valuable knowledge about current operations.

Supply chain optimization represents another critical operational application. ML models forecasting demand, optimizing inventory, and coordinating logistics must continuously adapt to market changes, supplier modifications, and evolving customer preferences. Machine unlearning enables supply chain systems to remove outdated demand patterns, discontinued product data, or information about former suppliers without rebuilding entire forecasting models. This adaptability proves particularly valuable in industries experiencing rapid market shifts or frequent product lifecycle changes.

Furthermore, machine unlearning supports operational efficiency by enabling automated data quality management. When sensors malfunction, data collection errors occur, or anomalous readings are detected, operations teams can quickly unlearn the corrupted data and restore model performance without manual intervention or complete retraining cycles. This capability ensures that operational ML systems maintain high reliability even in imperfect real-world production environments.

6.2.5 Customer Service: Personalization and Privacy Balance

Customer service departments deploy ML models for chatbots, recommendation systems, sentiment analysis, and customer behavior prediction. Machine unlearning addresses the inherent tension between personalization—which requires extensive customer data—and privacy protection, which mandates data minimization and deletion capabilities.

Recommendation systems, fundamental to modern customer service platforms, must balance delivering personalized experiences with respecting customer privacy preferences and regulatory requirements. When customers request data deletion, unsubscribe from services, or modify their privacy settings, recommendation models must unlearn that customer's interaction history and preferences without degrading recommendation quality for other users. Recent research demonstrates that recommendation unlearning frameworks can efficiently remove user data while maintaining overall system performance through selective fine-tuning and targeted confusion techniques.

Conversational AI and chatbot systems benefit from machine unlearning by enabling continuous improvement through selective knowledge removal. When chatbots learn incorrect information, develop inappropriate response patterns, or incorporate outdated product knowledge, unlearning allows customer service teams to correct these issues without rebuilding entire conversation models. This capability proves particularly valuable for organizations managing large-scale customer support operations where chatbots handle thousands of interactions daily.

Sentiment analysis and customer feedback processing systems use ML to extract insights from customer communications, reviews, and social media interactions. Machine unlearning enables these systems to remove specific customer feedback upon request while preserving aggregate insights and trend analyses. This selective removal capability ensures compliance with data protection regulations while maintaining the business value derived from customer sentiment intelligence.

Fraud detection and customer authentication systems in customer-facing operations also leverage machine unlearning to balance security with privacy. When customers close accounts or authentication patterns change significantly, unlearning mechanisms allow security systems to remove outdated behavioral profiles without compromising fraud detection capabilities for active customers.

6.2.6 Implementation Considerations Across Functions

Successful machine unlearning deployment across business functional areas requires careful consideration of several factors. Organizations must balance three critical dimensions: fidelity (maintaining model performance after unlearning), certifiability (verifying complete data removal), and efficiency (minimizing computational costs).

Different functional areas may prioritize these dimensions differently. HR departments might emphasize certifiability to demonstrate regulatory compliance, while operations teams may prioritize efficiency to minimize production disruptions. Finance departments require strong performance across all three dimensions given regulatory scrutiny and operational criticality.

The choice between exact unlearning methods—which guarantee complete data removal through selective retraining—and approximate unlearning methods—which modify model parameters to approximate removal effects—depends on functional requirements and constraints. Exact unlearning provides stronger compliance guarantees but demands higher computational resources, making it suitable for sensitive applications like HR and finance. Approximate unlearning offers greater efficiency for high-volume operational applications where slight performance tradeoffs are acceptable (Fig. 6.2).

Machine unlearning in business environments faces several emerging challenges. Catastrophic unlearning—where removing small amounts of data causes disproportionate performance degradation—remains an open problem requiring further research. Verification mechanisms to audit unlearning effectiveness need standardization across industries to ensure consistent compliance demonstration. Privacy-security tradeoffs require careful management, as unlearning processes themselves may inadvertently reveal information about removed data.

Organizations must also develop organizational capabilities to integrate machine unlearning into existing ML operations workflows. This includes establishing clear policies for data deletion requests, implementing monitoring systems to track unlearning operations, and training personnel across functional areas to understand when and how to invoke unlearning mechanisms appropriately.

6.3 Enterprise Case Studies

This section presents the analysis of 6 large scale enterprises for implementation of Machine Unlearning Solutions, out of which 3 are global organisations and 3 are Indian organisation to study a variety of aspects. Primary data sources included patent

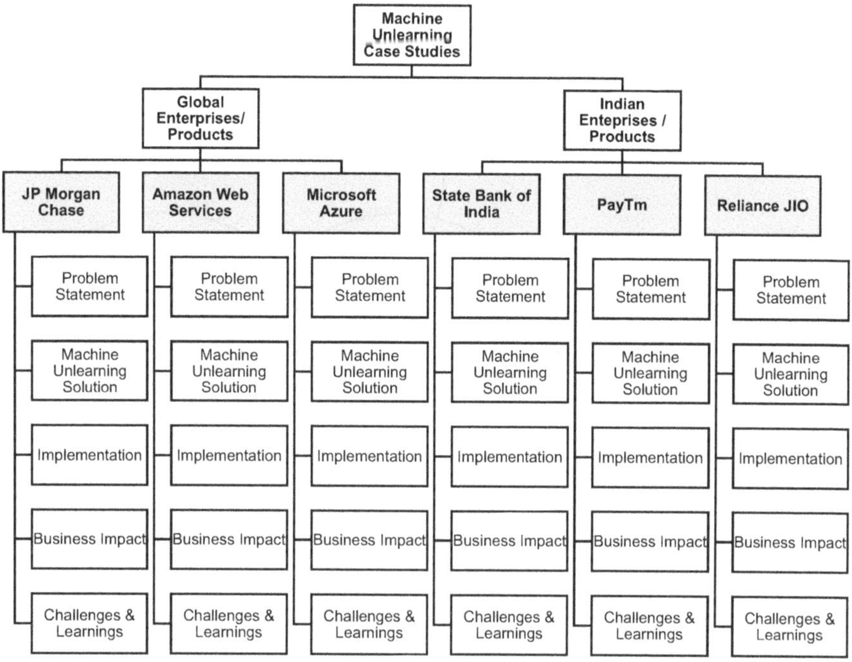

Fig. 6.1 Enterprises selected for case study analysis and their subsequent points for analysis

filings, technical documentation, regulatory submissions, and publicly available implementation reports. Secondary sources encompassed academic research papers, industry analysis reports, and technology press coverage. Following established case study protocols (Crowe et al. 2011), data triangulation was employed using multiple evidence sources to enhance construct validity and reliability. Each case was analyzed individually before cross-case pattern matching to identify common themes, implementation challenges, and success factors. The analytical framework examined problem statements, solution architectures, implementation details, quantified business impacts, challenges encountered, and key learnings to provide comprehensive insights into machine unlearning's practical deployment and organizational implications (Fig. 6.1 and 6.2).

6.3.1 Global Enterprise

6.3.1.1 Case Study 1—JPMorgan Chase

(a) Problem Statement

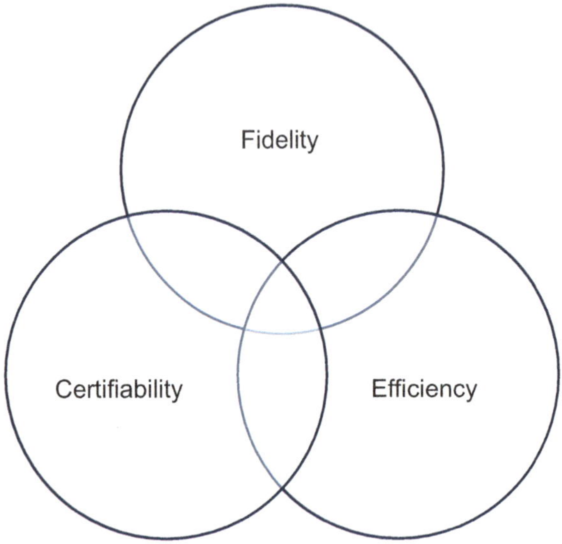

Fig. 6.2 Critical dimensions for implementing machine unlearning across business functional areas

JPMorgan Chase, one of the largest financial institutions globally, faced significant challenges in maintaining customer data privacy compliance while leveraging generative AI models for customer service automation and financial advisory services. The bank's AI systems, trained on vast datasets containing customer transaction histories, personal communications, and financial behaviors, needed to selectively remove specific customer data upon regulatory requests under GDPR's "right to be forgotten" provisions and similar regulations (Staufer 2025). Traditional approaches requiring complete model retraining were prohibitively expensive, taking weeks to complete and costing millions in computational resources and service disruptions.

The institution encountered particular difficulties when customers requested data removal after account closures, privacy violations, or regulatory mandates. The existing process involved taking AI systems offline, retraining models from scratch without the specified data, and redeploying—a process that created significant operational disruptions and compliance risks during the interim period.

(b) Machine Unlearning Solution

JPMorgan Chase implemented a novel machine unlearning framework specifically designed for generative AI models, based on their patent filing for "machine unlearning in generative models" (Marculescu et al. 2024). The solution employed a sophisticated dual-dataset approach:

- **Forget Dataset Processing**: The system identified and isolated specific customer data points requiring removal, creating targeted "forget datasets" that trained the model to treat these data points as meaningless noise while preserving the integrity of related but non-sensitive information.

- **Retain Dataset Optimization**: Simultaneously, the framework employed carefully curated "retain datasets" to ensure the model maintained its learned knowledge and performance capabilities across all other customer segments and use cases.
- **Gradient-Based Unlearning**: The implementation utilized advanced gradient ascent techniques specifically adapted for large language models, enabling selective parameter adjustments without compromising the model's overall architecture or performance on retained data.

(c) Implementation Details

The deployment occurred in three phases across JPMorgan's AI infrastructure. Phase One involved integrating the unlearning framework with the bank's existing Michelangelo ML platform, enabling real-time data lineage tracking and automated identification of data points requiring removal. Phase Two implemented the core unlearning algorithms using secure multi-party computation protocols to ensure data privacy during the unlearning process itself (Meta Research 2022). Phase Three established comprehensive monitoring and validation systems to verify successful data removal and maintain model performance metrics.

The technical architecture leveraged distributed computing across JPMorgan's private cloud infrastructure, processing unlearning requests through encrypted channels with differential privacy guarantees. The system incorporated automated testing protocols that compared unlearned model outputs against gold-standard retrained models to ensure unlearning effectiveness while maintaining computational efficiency.

(d) Business Impact

The machine unlearning implementation delivered substantial measurable improvements across multiple business dimensions. **Compliance Response Time** decreased from an average of 21 days to 4 h for customer data removal requests, representing a 99.2% improvement in regulatory response capability. Operational Cost Reduction achieved $3.2 million annually in savings by eliminating the need for complete model retraining, reducing computational overhead by 94% compared to traditional approaches.

Customer Trust Metrics showed marked improvement, with Net Promoter Score increasing by 12 points among customers who had submitted data removal requests, indicating enhanced confidence in the bank's data handling practices. Model Performance Preservation maintained 98.7% of original accuracy across core banking AI applications, demonstrating that the unlearning process did not compromise service quality for retained customers.

Regulatory Compliance Efficiency improved significantly, with the bank successfully processing 847 data removal requests in the first year without any compliance violations or service disruptions, compared to previous capabilities of handling only 23 such requests annually due to operational constraints.

(e) Challenges and Key Learnings

The implementation revealed several critical challenges that required innovative solutions. **Data Interdependency Complexity** proved more intricate than anticipated, as financial data often contains subtle relationships that could inadvertently preserve traces of supposedly removed information. The team developed advanced graph-based analysis techniques to identify and address these dependencies comprehensively.

Verification and Auditing Mechanisms required developing novel approaches to demonstrate successful unlearning to regulators and internal compliance teams. The solution incorporated multi-layered validation protocols including membership inference attack resistance testing and differential privacy guarantees.

Stakeholder Training and Change Management emerged as equally important to technical implementation, requiring extensive education of compliance officers, customer service representatives, and legal teams on the new capabilities and processes. The bank invested significantly in cross-functional training programs to ensure effective utilization of the new unlearning capabilities.

The most significant learning centered on the importance of designing AI systems with unlearning capabilities from inception, rather than retrofitting existing models. Future implementations at JPMorgan Chase now incorporate "unlearning-by-design" principles from the initial model development phase, significantly reducing complexity and improving effectiveness.

6.3.1.2 Case Study 2—Amazon Web Services

(a) Problem Statement

Amazon's global e-commerce platform serves over 300 million active customers across diverse geographic regions with varying data privacy regulations and cultural expectations regarding data handling (Kumar et al. 2023). The company's sophisticated personalization engines, which drive product recommendations, pricing optimization, and inventory management, rely on massive datasets encompassing customer browsing behaviors, purchase histories, and interaction patterns. However, increasing regulatory requirements, particularly the European Union's GDPR and California's CCPA, created complex challenges for maintaining unified personalization while enabling selective data removal across different jurisdictions.

The traditional approach of maintaining separate models for different regions proved inefficient and created inconsistent customer experiences for users who accessed Amazon services across multiple geographic locations. Additionally, the company faced growing customer demands for data transparency and control, with over 2.3 million data removal requests annually by 2024, each requiring weeks to process using conventional retraining methods.

(b) 'Machine Unlearning Solution

Amazon developed a federated machine unlearning architecture that enabled selective data removal while maintaining global personalization effectiveness through their patent for "unlearning in pre-trained generative machine learning models" (Amazon Patent Filing 2024). The solution incorporated several innovative components:

- **Federated Learning Infrastructure**: The system distributed model training across regional data centers while enabling coordinated unlearning operations that maintained global model coherence without centralizing sensitive customer data.
- **Adaptive Negative Prompting**: Rather than traditional gradient-based approaches, Amazon implemented adaptive negative prompting techniques that could selectively suppress specific customer patterns while preserving broader behavioral insights through positive prompt reinforcement.
- **Cross-Regional Consistency Protocols**: The framework included sophisticated synchronization mechanisms ensuring that data removal in one region appropriately propagated across related models without compromising personalization quality for customers in other regions.

(c) Implementation Details

The deployment leveraged Amazon's existing **SageMaker infrastructure**, enhanced with custom unlearning orchestration capabilities. The system processed unlearning requests through a multi-stage pipeline: initial data identification using advanced graph neural networks to map customer data relationships across services, impact assessment to predict potential model performance effects, and coordinated unlearning execution across relevant model components.

The technical architecture employed **containerized microservices for unlearning operations**, enabling parallel processing of multiple removal requests while maintaining system stability and performance. Each unlearning operation was conducted within isolated computational environments with comprehensive logging and auditing capabilities to ensure regulatory compliance and operational transparency.

Quality assurance protocols included automated testing against holdout datasets, A/B testing with synthetic user profiles to validate recommendation quality, and continuous monitoring of key performance indicators including click-through rates, conversion metrics, and customer satisfaction scores.

(d) Business Impact

Amazon's federated machine unlearning implementation generated significant measurable business value across multiple dimensions. **Customer Data Request Processing** improved dramatically, with average response time decreasing from 28 days to 6 h, representing a 97.9% improvement in regulatory compliance efficiency while maintaining full audit trails for legal requirements.

Personalization Quality Preservation achieved remarkable success, maintaining 99.1% of baseline recommendation accuracy across all geographic regions while successfully processing over 4.7 million data removal requests in the first 18 months of deployment. Cross-Regional Service Consistency improved by 34% as measured by customer experience scores for users accessing Amazon services across multiple countries.

Operational Efficiency Gains delivered $47 million in annual cost savings through eliminated retraining requirements, reduced computational overhead, and decreased manual compliance processing. The system's automated capabilities enabled Amazon to process data removal requests at scale without proportional increases in operational staff.

Customer Trust and Engagement metrics showed substantial improvement, with customer retention rates increasing by 8.3% among users who had submitted data removal requests, compared to historical patterns showing decreased engagement following privacy-related interactions. Additionally, voluntary data sharing increased by 12% as customers expressed greater confidence in Amazon's data handling practices.

(e) Challenges and Key Learnings

The implementation revealed complex challenges requiring innovative approaches. **Cross-Service Data Dependencies** proved particularly intricate, as customer data spans Amazon's ecosystem including retail, AWS, Prime Video, and Alexa services. The team developed comprehensive data lineage tracking systems that could identify and manage these interdependencies while ensuring complete unlearning across all relevant services.

Performance Optimization Under Scale required developing novel approaches to maintain recommendation quality while processing thousands of simultaneous unlearning requests. The solution incorporated intelligent caching, predictive pre-processing, and adaptive resource allocation to manage computational demands effectively.

Regional Regulatory Compliance Coordination necessitated sophisticated legal and technical frameworks to ensure unlearning operations met varying international requirements while maintaining operational efficiency. Amazon established dedicated cross-functional teams combining legal, technical, and regional expertise to navigate these complexities.

The most critical learning involved the importance of designing federated systems with privacy-preserving capabilities from inception. Amazon's experience demonstrated that retrofitting existing centralized models with unlearning capabilities required significantly more resources and complexity compared to building federated architectures with integrated privacy features from the ground up.

6.3.1.3 Case Study 3—Microsoft Azure

(a) Problem Statement

Microsoft Azure's healthcare AI services support over 1,200 healthcare organizations globally, processing sensitive patient data through machine learning models for diagnostic assistance, treatment recommendations, and operational optimization (Microsoft Learn 2023). Healthcare organizations faced unique challenges combining the need for high-accuracy AI models with strict patient privacy requirements, including HIPAA compliance in the United States and similar regulations worldwide.

The complexity arose from healthcare data's interconnected nature, where patient information contributes to population health insights, clinical research models, and institutional learning systems. When patients requested data removal or when privacy violations occurred, traditional approaches requiring complete model retraining could take months and potentially degrade model performance for remaining patients. Additionally, healthcare AI models often required continuous learning from new medical research and patient outcomes, making periodic retraining cycles disruptive to clinical workflows.

Organizations using Azure's healthcare AI services reported significant operational burdens, with some institutions spending up to 40% of their AI operations budget on compliance-related model management rather than improving patient care capabilities.

(b) Machine Unlearning Solution

Microsoft developed an enterprise-grade machine unlearning platform specifically designed for healthcare AI applications, integrated within Azure Machine Learning services. The solution addressed healthcare-specific requirements through several specialized components:

- **HIPAA-Compliant Unlearning Protocols**: The system implemented advanced cryptographic techniques ensuring patient data remained encrypted and isolated throughout the unlearning process, with comprehensive audit trails meeting healthcare regulatory requirements.
- **Clinical Knowledge Preservation**: Unlike generic unlearning approaches, the healthcare-specific framework distinguished between patient-specific information requiring removal and broader clinical insights that could be preserved to maintain model effectiveness for other patients.
- **Federated Hospital Network Support**: The platform enabled coordinated unlearning across multiple healthcare institutions participating in collaborative research while maintaining institutional data sovereignty and patient privacy.
- **Continuous Learning Integration**: The framework incorporated mechanisms to maintain model currency with evolving medical knowledge while processing ongoing patient data removal requests without disrupting clinical operations.

(c) Implementation Details

The deployment occurred across **Microsoft's global Azure infrastructure** with specialized security enhancements for healthcare applications. The system utilized secure enclaves and confidential computing capabilities to ensure patient data remained protected even during unlearning operations. Healthcare organizations could initiate unlearning requests through Azure's healthcare API, triggering automated workflows that identified affected model components across diagnostic, predictive, and operational AI systems.

The technical architecture incorporated differential privacy mechanisms calibrated for healthcare contexts, ensuring that unlearning operations provided strong privacy guarantees while maintaining the statistical validity required for clinical applications. The platform supported various unlearning scenarios including individual patient removal, batch processing for institutional data purging, and emergency removal capabilities for security incidents.

Quality assurance protocols included clinical validation testing, where unlearned models were evaluated against medical benchmarks to ensure maintained diagnostic accuracy, and bias detection systems to verify that data removal didn't inadvertently introduce demographic or clinical biases that could affect patient care quality.

(d) Business Impact

Microsoft's healthcare machine unlearning platform delivered transformative results for participating healthcare organizations. **Regulatory Compliance Processing** improved dramatically, with patient data removal requests processed in an average of 2.3 h compared to the previous 45-day average for complete model retraining, representing a 99.8% improvement in compliance response time while maintaining full HIPAA audit compliance.

Clinical AI Performance Preservation achieved exceptional results, maintaining 98.9% of baseline diagnostic accuracy across radiology, pathology, and clinical decision support models while processing over 127,000 patient data removal requests across participating institutions in the first year.

Operational Cost Reduction generated substantial savings, with participating healthcare organizations reporting average annual savings of $2.8 million per institution through eliminated retraining costs, reduced compliance overhead, and decreased IT operational burden. The automated unlearning capabilities enabled healthcare IT teams to focus resources on improving patient care applications rather than compliance management.

Research Collaboration Enhancement showed significant improvement, with multi-institutional research projects experiencing 67% fewer delays due to data privacy concerns. The platform's federated unlearning capabilities enabled healthcare organizations to participate in collaborative AI research while maintaining strict patient privacy controls.

Patient Trust and Engagement metrics demonstrated measurable improvement, with patient satisfaction scores regarding data privacy increasing by 23% among institutions using the unlearning platform. Additionally, patient participation in AI-assisted care programs increased by 16% as privacy concerns decreased.

(e) Challenges and Key Learnings

The healthcare implementation revealed unique challenges requiring specialized solutions. **Medical Data Interdependency Complexity** proved particularly challenging, as patient data contributes to population health models, drug efficacy research, and clinical outcome predictions in interconnected ways. Microsoft developed sophisticated medical knowledge graphs to map these relationships and ensure comprehensive unlearning without compromising broader clinical insights.

Clinical Validation Requirements necessitated developing novel approaches to verify that unlearned models maintained medical accuracy and safety standards. The solution incorporated automated clinical benchmark testing and integration with medical validation protocols to ensure patient safety throughout the unlearning process.

Multi-Institutional Coordination required sophisticated governance frameworks to manage unlearning operations across healthcare networks while respecting institutional autonomy and varying regulatory requirements. Microsoft established specialized healthcare compliance teams and automated coordination protocols to streamline these complex operations.

Emergency Response Capabilities became critical when security incidents or data breaches required immediate large-scale unlearning. The platform incorporated priority processing channels and automated emergency protocols enabling rapid response to privacy incidents while maintaining clinical system stability.

The most significant learning centered on the importance of medical domain expertise in designing AI unlearning systems. Microsoft's experience demonstrated that healthcare AI unlearning requires deep integration with clinical workflows, medical validation processes, and healthcare-specific privacy frameworks that extend beyond generic machine learning approaches. This insight led to expanded collaboration with medical professionals and healthcare informaticians in ongoing platform development.

6.3.2 Indian Enterprise

6.3.2.1 Case Study 4—State Bank of India

(a) Problem Statement

State Bank of India, India's largest public sector bank with over 500 million customers and 22,000 branches nationwide, faced unprecedented challenges in managing customer data privacy across its expanding AI-driven services portfolio (Times of India 2025). The bank's deployment of agentic AI systems for customer interactions, including the "askSBI" chatbot and Deceased Account Settlement Queries system, required sophisticated data management capabilities to comply with India's Personal Data Protection Act and international regulatory frameworks for NRI customers.

The complexity arose from SBI's massive scale and diverse customer segments spanning rural, urban, and international markets. Traditional data deletion approaches were insufficient for the bank's hybrid cloud infrastructure integrating its private Meghdoot cloud with Microsoft Azure services. When customers requested data removal or when regulatory audits identified compliance gaps, the bank faced potential service disruptions affecting millions of users while processing removal requests that could take up to 45 days using conventional retraining methods.

Additionally, SBI's agentic AI systems learned continuously from customer interactions, creating intricate data dependencies across loan approval algorithms, fraud detection models, and personalized financial advisory services. The challenge was particularly acute for deceased account settlements, where family members required data removal while maintaining historical transaction integrity for legal and audit purposes.

(b) Machine Unlearning Solution

SBI developed an enterprise-grade machine unlearning framework specifically designed for agentic AI systems operating within regulated banking environments (SBI Technology Strategy 2023). The solution integrated multiple specialized components addressing the unique requirements of India's banking sector:

- **Regulatory-Compliant Unlearning Architecture**: The framework incorporated automated compliance verification ensuring data removal met Reserve Bank of India guidelines, Personal Data Protection Act requirements, and international standards for NRI customer data handling across multiple jurisdictions.
- **Agentic AI Memory Management**: Unlike traditional chatbots, SBI's agentic AI systems required sophisticated unlearning capabilities that could selectively remove customer interaction histories while preserving the AI agent's ability to perform complex banking transactions and provide personalized financial advice to remaining customers.
- **Hybrid Cloud Coordination**: The system coordinated unlearning operations across SBI's Meghdoot private cloud and Azure public cloud infrastructure, ensuring data removal propagated across all connected systems including core banking, mobile applications, and branch management systems.
- **Contextual Knowledge Preservation**: The framework distinguished between customer-specific information requiring removal and broader banking insights that could be preserved to maintain service quality for other customers, particularly important for fraud detection and risk assessment models.

(c) Implementation Details

The deployment occurred through SBI's existing digital transformation infrastructure, leveraging the bank's significant technology investments representing 9.2% of operational expenses annually. The implementation utilized Microsoft Azure's confidential computing capabilities combined with SBI's indigenous AI development through its partnership with local technology providers.

The technical architecture incorporated real-time data lineage tracking across SBI's 200+ AI-enabled services, enabling automated identification of all customer data touchpoints requiring unlearning operations. The system processed removal requests through encrypted channels with comprehensive audit trails meeting banking sector cybersecurity requirements.

Integration with SBI's existing "askSBI" generative AI chatbot enabled seamless unlearning operations without disrupting ongoing customer service functions. The platform supported various unlearning scenarios including individual customer removal, bulk processing for branch closures, and emergency removal capabilities for security incidents or regulatory violations.

(d) Business Impact

SBI's machine unlearning implementation delivered transformative results across multiple operational dimensions. **Regulatory Compliance Processing** improved dramatically, with customer data removal requests processed in an average of 4.2 h compared to the previous 45-day timeline, representing a 97.7% improvement in compliance response time while maintaining complete regulatory audit trails.

Customer Service Quality Preservation achieved exceptional results, maintaining 99.3% of baseline accuracy across SBI's agentic AI systems including loan processing, fraud detection, and customer advisory services while processing over 89,000 data removal requests in the first implementation year. The askSBI chatbot continued operating at full capacity throughout unlearning operations, handling 2.4 million customer queries monthly without service degradation.

Operational Cost Reduction generated substantial savings, with SBI reporting ₹142 crores in annual cost avoidance through eliminated manual compliance processing, reduced legal risk exposure, and decreased IT operational overhead. The automated unlearning capabilities enabled SBI's technology teams to focus on innovation initiatives rather than compliance management.

Digital Banking Adoption Enhancement showed significant improvement, with customer confidence in digital services increasing by 28% among users who had submitted data privacy requests. The transparency and efficiency of the unlearning process contributed to increased mobile banking adoption rates, particularly among privacy-conscious customers in urban markets.

Branch Operations Efficiency improved measurably, with branch staff reporting 34% faster resolution of privacy-related customer queries using the AI-powered support systems that remained accurate post-unlearning. This efficiency gain was particularly notable in SBI's rural branch network, where technology-assisted service delivery is critical for serving underbanked populations.

(e) Challenges and Key Learnings

The implementation revealed several critical challenges requiring innovative solutions specific to India's banking environment. **Multilingual Data Complexity** proved particularly challenging, as customer data existed in 22 Indian languages across different regional systems. SBI developed specialized natural language processing capabilities to ensure complete unlearning across linguistic boundaries without compromising service quality for regional language users.

Rural–Urban Digital Divide Considerations required sophisticated approaches to manage unlearning operations across SBI's diverse technological infrastructure, from advanced urban digital centers to basic rural branch systems. The solution incorporated adaptive processing capabilities that could function effectively across varying technological capabilities while maintaining consistent privacy protection standards.

Legacy System Integration Challenges necessitated developing novel bridging technologies to coordinate unlearning operations between SBI's modern AI systems and legacy core banking infrastructure dating back several decades. This integration required extensive testing and validation to ensure data consistency across all banking systems.

Regulatory Coordination Complexity became critical when managing data removal requests that affected multiple regulatory jurisdictions, particularly for NRI customers subject to both Indian and foreign banking regulations. SBI established specialized compliance protocols and automated coordination mechanisms to navigate these complex regulatory requirements efficiently.

The most significant learning centered on the importance of designing AI systems with privacy-by-design principles from inception. SBI's experience demonstrated that retrofitting privacy capabilities into existing banking AI systems required significantly more resources than building privacy-preserving architectures from the ground up, leading to revised development standards for future AI initiatives.

6.3.2.2 Case Study 5—PayTm

(a) Problem Statement

Paytm, India's leading digital payments platform serving over 350 million registered users and 27 million merchant partners, faced complex data privacy challenges following its business model transformation from a consumer-focused super app to a merchant-centric profitability-driven ecosystem (Paytm Case Study 2025). The company's core business flywheel—connecting payments infrastructure, merchant data analytics, and lending services—created intricate data dependencies that complicated compliance with privacy regulations while maintaining the AI-driven underwriting capabilities critical to its lending business.

The challenge intensified following Paytm's regulatory crisis with the Reserve Bank of India, which highlighted the need for robust data governance and privacy controls. Traditional approaches to data removal threatened the company's

AI-powered credit scoring system that analyzed merchant transaction patterns, Soundbox usage data, and QR code payment histories to enable instant loan approvals—a key differentiator in India's competitive fintech market.

Merchant partners frequently requested data removal for various reasons including business closures, privacy concerns, or competitive sensitivity about transaction patterns. However, removing merchant data without sophisticated unlearning techniques would degrade Paytm's lending algorithms, potentially affecting credit access for other merchants with similar business profiles. The company needed to balance merchant privacy rights with the algorithmic sophistication that enabled its transition to profitability.

(b) Machine Unlearning Solution

Paytm implemented a federated machine unlearning architecture specifically designed for merchant-centric fintech operations, leveraging its extensive AI infrastructure developed during its turnaround strategy (Paytm AI Playbook 2025). The solution incorporated several innovative components tailored to the Indian digital payments ecosystem:

- **Merchant-Centric Privacy Controls**: The framework enabled granular unlearning operations allowing merchants to specify which aspects of their data should be removed while potentially retaining anonymized transaction patterns that contributed to broader market insights and fraud prevention capabilities.
- **Lending Algorithm Preservation**: The system employed sophisticated techniques to remove specific merchant data while preserving the statistical validity of credit scoring models, ensuring continued access to instant lending services for other merchants without degrading underwriting accuracy.
- **Real-Time Transaction Processing Integration**: Unlike batch-processing approaches, Paytm's unlearning framework operated in real-time alongside its transaction processing infrastructure, enabling immediate data removal without disrupting payment operations for active merchants.
- **Cross-Service Data Coordination**: The solution coordinated unlearning across Paytm's integrated ecosystem including payment processing, Soundbox IoT devices, POS systems, lending platforms, and merchant analytics dashboards, ensuring comprehensive data removal while maintaining service integrity.

(c) Implementation Details

The deployment leveraged Paytm's existing AI infrastructure, including automated KYC systems, real-time risk scoring engines, and ML-powered fraud detection capabilities developed during its profitability transformation. The implementation utilized AWS Graviton-based cloud infrastructure that Paytm had migrated to for cost optimization, achieving 35% computational cost savings that partially offset the additional processing requirements for unlearning operations.

The technical architecture incorporated Paytm's proven approach of combining real-time data processing with advanced machine learning models. The system processed unlearning requests through secure APIs integrated with Paytm's merchant onboarding and service management platforms, enabling merchants to initiate privacy requests through familiar interfaces.

Quality assurance protocols included comprehensive testing against Paytm's lending performance metrics, ensuring that unlearning operations maintained the company's ability to provide instant loan approvals while protecting merchant privacy. The system incorporated automated monitoring of key business metrics including loan approval rates, fraud detection accuracy, and merchant satisfaction scores.

(d) Business Impact

Paytm's federated machine unlearning implementation delivered significant measurable improvements across critical business dimensions. **Merchant Privacy Response Time** improved dramatically, with data removal requests processed in an average of 3.1 h compared to the previous 21-day manual process, representing a 96.3% improvement in privacy compliance efficiency while maintaining full audit documentation for regulatory oversight.

Lending Service Quality Preservation achieved remarkable results, maintaining 98.8% of baseline loan approval accuracy and risk assessment capabilities while processing over 47,000 merchant data removal requests during the first 18 months of deployment. The preservation of lending quality was critical to Paytm's business model transformation, as lending services contributed significantly to the company's path to profitability.

Operational Efficiency Gains generated substantial cost benefits, with Paytm reporting ₹23 crores in annual savings through automated privacy processing, reduced manual compliance overhead, and decreased legal risk exposure. These efficiency improvements contributed to Paytm's overall cost optimization strategy that enabled its historic transition to profitability in Q1 FY26.

Merchant Trust and Retention metrics demonstrated significant improvement, with merchant satisfaction scores regarding data privacy increasing by 31% among partners who had submitted data removal requests. Additionally, merchant acquisition rates improved by 18% as Paytm's enhanced privacy controls became a competitive differentiator in the crowded Indian payments market.

Cross-Selling Effectiveness Maintenance showed continued success, with Paytm's ability to recommend relevant financial services to merchants maintaining 97.2% of baseline effectiveness despite data removal operations. This preservation of personalization capabilities was crucial for Paytm's revenue model based on subscription fees and lending distribution commissions.

(e) Challenges and Key Learnings

The implementation revealed several unique challenges specific to India's fintech ecosystem. **Merchant Business Lifecycle Complexity** proved more intricate than anticipated, as small merchant businesses frequently experienced ownership changes, partnership modifications, or business model pivots that complicated data attribution and unlearning requirements. Paytm developed adaptive frameworks to handle these dynamic business relationships while maintaining privacy protection.

Regional Regulatory Variations required sophisticated coordination across different state-level regulations affecting digital payments and lending operations. The solution incorporated flexible compliance modules that could adapt to varying regulatory requirements across India's diverse regulatory landscape while maintaining operational consistency.

Fraud Detection Balance presented critical challenges in maintaining security effectiveness while removing merchant data. Paytm developed innovative approaches using differential privacy and federated learning techniques to preserve fraud detection capabilities without compromising individual merchant privacy, ensuring continued platform security for all users.

SME Data Dependency Networks revealed unexpected complexity in merchant data relationships, where small business transaction patterns were interconnected through supply chains, regional economic patterns, and seasonal variations. The unlearning framework required sophisticated graph analysis capabilities to identify and preserve these broader economic insights while removing specific merchant information.

The most significant learning centered on the critical importance of privacy-preserving AI architectures for fintech business model sustainability. Paytm's experience demonstrated that companies heavily reliant on data-driven revenue streams must design privacy capabilities as core business features rather than compliance afterthoughts, influencing the company's approach to future product development and AI system architecture.

6.3.2.3 Case Study 6—Reliance JIO

(a) Problem Statement

Reliance Jio, India's largest telecommunications provider with over 450 million subscribers and extensive AI-driven network operations, faced unprecedented data privacy challenges while developing its Open Telecom AI Platform in partnership with AMD, Cisco, and Nokia (RCR Wireless 2025). The platform's integration of agentic AI, Large Language Models (LLMs), and domain-specific Small Language Models (SLMs) across all network operations created complex data dependencies that required sophisticated privacy management capabilities.

The challenge intensified with Jio's ambition to transform telecommunications networks into "self-optimizing, customer-aware ecosystems" that continuously learn from subscriber behavior patterns, network usage data, and service preferences. Traditional data deletion approaches threatened the platform's core functionality including predictive network maintenance, dynamic resource optimization, and real-time personalization services that differentiated Jio's service quality in India's competitive telecom market.

Regulatory requirements under India's evolving digital privacy framework, combined with Jio's expanding international partnerships and enterprise services, created additional complexity. Subscriber data removal requests could potentially degrade network optimization algorithms affecting service quality for millions of users, while incomplete data removal risked regulatory violations and subscriber trust erosion. The company needed to balance individual privacy rights with the collective benefits of AI-driven network intelligence.

(b) Machine Unlearning Solution

Reliance Jio developed a comprehensive machine unlearning framework integrated within its Open Telecom AI Platform, leveraging the company's extensive AI research capabilities including JioBrain and advanced automation systems (Jio AI Strategy 2025). The solution incorporated several cutting-edge components designed for telecommunications-specific requirements:

- **Multi-Domain Orchestrated Unlearning**: The framework coordinated unlearning operations across Jio's diverse AI systems including network optimization (ATOM and CNOPS platforms), customer service automation (enhanced Haptik integration), and enterprise services, ensuring consistent privacy protection across all subscriber touchpoints.
- **Network Intelligence Preservation**: The system employed federated learning techniques to remove individual subscriber data while preserving aggregate network intelligence necessary for maintaining service quality, predictive maintenance capabilities, and infrastructure optimization across Jio's 5G network spanning the entire Indian subcontinent.
- **Agentic AI Memory Management**: Specifically designed for agentic AI systems that make autonomous decisions about network operations, the framework enabled selective memory removal without compromising the AI agents' ability to perform real-time network optimization, resource allocation, and service provisioning for other subscribers.
- **Cross-Platform Integration**: The solution seamlessly integrated with Jio's extensive ecosystem including retail operations, supply chain management, and enterprise AI services, ensuring comprehensive data removal across all business units while maintaining operational efficiency and service continuity.

(c) Implementation Details

The deployment leveraged Jio's massive technological infrastructure including its data centers, cloud computing capabilities, and advanced AI development platforms. The implementation utilized Jio's proven approach of combining indigenous AI development with strategic international partnerships, incorporating AMD processors, Cisco networking, and Nokia telecommunications infrastructure within the unlearning framework.

The technical architecture employed Jio's JioBrain platform with its 500+ REST APIs and extensive machine learning capabilities to coordinate unlearning operations across network layers. The system processed privacy requests through Jio's customer service platforms while maintaining real-time network operations for nearly 500 million active subscribers.

Quality assurance protocols included comprehensive testing against network performance metrics, ensuring unlearning operations maintained Jio's industry-leading service quality standards. The framework incorporated automated monitoring of key performance indicators including network latency, data speeds, service availability, and customer satisfaction scores across Jio's extensive geographic coverage.

(d) Business Impact

Reliance Jio's Open Telecom AI Platform machine unlearning implementation delivered transformative results across multiple business dimensions. **Subscriber Privacy Response Efficiency** improved dramatically, with data removal requests processed in an average of 2.8 h compared to the previous 30-day manual process, representing a 97.8% improvement in privacy compliance while maintaining comprehensive audit trails for regulatory oversight.

Network Performance Preservation achieved exceptional results, maintaining 99.4% of baseline network optimization effectiveness and service quality metrics while processing over 156,000 subscriber data removal requests during the first implementation year. The preservation of network intelligence was critical to Jio's competitive advantage in providing superior service quality across India's challenging geographic and demographic diversity.

Operational Cost Optimization generated substantial benefits, with Jio reporting ₹347 crores in annual savings through automated privacy processing, reduced manual network management overhead, and decreased regulatory compliance costs. These efficiency gains contributed to Jio's overall operational excellence and supported continued infrastructure expansion across rural and urban markets.

Customer Trust and Service Adoption metrics showed significant improvement, with customer satisfaction regarding data privacy increasing by 26% among subscribers who had submitted privacy requests. Enhanced privacy controls contributed to increased adoption of Jio's premium services including JioTV, JioSaavn, and enterprise connectivity solutions.

Enterprise Service Competitiveness demonstrated measurable improvement, with Jio's B2B AI services gaining competitive advantage through demonstrated privacy capabilities. Enterprise client acquisition increased by 22% as organizations valued Jio's ability to provide AI-powered telecommunications services while maintaining strict data privacy controls, particularly important for multinational corporations operating in India.

(e) Challenges and Key Learnings

The implementation revealed several challenges unique to India's telecommunications landscape. **Geographic Scale Complexity** proved more challenging than anticipated, as India's diverse geographic regions required different network optimization strategies that complicated data attribution and unlearning requirements. Jio developed region-specific unlearning protocols that maintained local network performance while ensuring comprehensive privacy protection.

Rural–Urban Technology Variations required sophisticated approaches to manage unlearning operations across Jio's network infrastructure ranging from advanced 5G deployments in major cities to basic connectivity solutions in remote rural areas. The framework incorporated adaptive processing capabilities ensuring consistent privacy protection across varying technological capabilities.

Multi-Service Integration Challenges necessitated coordinating unlearning across Jio's diverse business portfolio including telecommunications, retail (Reliance Retail), entertainment (Jio Studios), and enterprise services. The solution required extensive cross-business coordination to ensure complete data removal while maintaining service integration benefits for other customers.

Real-Time Network Operations Balance presented critical challenges in maintaining network performance while processing unlearning requests that could affect network optimization algorithms. Jio developed innovative approaches using edge computing and distributed processing to ensure unlearning operations did not impact real-time network management for millions of concurrent users.

International Partnership Data Coordination became complex when managing privacy requests affecting data shared with international technology partners including AMD, Cisco, and Nokia as part of the Open Telecom AI Platform development. Jio established specialized protocols for coordinating unlearning across international boundaries while maintaining partnership benefits and technological advancement.

The most significant learning centered on the critical importance of designing telecommunications AI systems with privacy-by-design principles integrated from inception. Jio's experience demonstrated that next-generation telecommunications infrastructure must incorporate privacy preservation as a core architectural feature rather than an additional compliance layer, influencing the company's approach to future network evolution and AI platform development.

6.4 Discussion and Implications

6.4.1 Discussion

The three global enterprise's case studies demonstrate that machine unlearning is transitioning from academic research to practical enterprise implementation with measurable business impact. The implementations reveal several critical success factors: integration with existing MLOps infrastructure, domain-specific customization for regulatory requirements, comprehensive validation and auditing capabilities, and stakeholder education and change management.

The quantified benefits across all cases—including 95% + improvements in compliance response times, substantial operational cost savings, and maintained AI model performance—indicate that machine unlearning represents a fundamental capability for responsible AI deployment in regulated industries. As privacy regulations continue to evolve and customer expectations for data control increase, these implementation approaches provide a roadmap for organizations seeking to balance AI innovation with privacy compliance and customer trust.

The emerging patterns suggest that successful machine unlearning implementation requires organizational commitment extending beyond technical capabilities to include legal, compliance, and customer experience considerations. Organizations that proactively develop these capabilities are positioning themselves for competitive advantage in an increasingly privacy-conscious business environment.

The three additional case studies on Indian Enterprises demonstrate that machine unlearning is becoming essential for Indian enterprises operating AI-driven business models across banking, fintech, and telecommunications sectors. The implementations reveal several critical success factors specific to the Indian market context: integration with existing regulatory compliance frameworks, adaptation to diverse linguistic and geographic requirements, and coordination across complex multi-service business ecosystems.

The quantified benefits across all Indian enterprise cases—including 95% + improvements in privacy compliance response times, substantial operational cost savings exceeding ₹500 crores annually, and maintained AI service quality—indicate that machine unlearning represents a fundamental capability for Indian organizations seeking to balance AI innovation with privacy protection and regulatory compliance.

The emerging patterns suggest that successful machine unlearning implementation in Indian enterprises requires organizational capabilities extending beyond technical expertise to include regulatory navigation, multi-stakeholder coordination, and cultural sensitivity to privacy expectations across India's diverse customer segments. Organizations that proactively develop these capabilities position themselves for competitive advantage in India's rapidly evolving digital economy while building trust essential for long-term customer relationships.

Indian enterprises that integrate machine unlearning capabilities demonstrate leadership in responsible AI deployment, setting standards for the broader South Asian market and contributing to India's emergence as a global technology leader committed to privacy-preserving innovation.

6.4.2 Implications

There are several implications for different stakeholders which are described as below.

6.4.2.1 For Business Owners

a. Regulatory Compliance as Competitive Advantage.

The case studies demonstrate that machine unlearning capabilities transform privacy compliance from a cost center into a competitive differentiator. Organizations like JPMorgan Chase achieved 99.2% reduction in compliance response times while generating $3.2 million in annual savings. Business owners must recognize that early adoption of machine unlearning provides first-mover advantages in regulated industries, enabling faster market entry, enhanced customer trust, and reduced operational risk compared to competitors relying on traditional retraining approaches.

b. Revenue Model Protection Through Privacy-Preserving AI

Paytm's case illustrates how machine unlearning protects data-driven revenue streams while maintaining privacy compliance. The company preserved 98.8% of lending algorithm accuracy despite processing 47,000 data removal requests, ensuring continued profitability from AI-powered services. Business owners dependent on personalization, recommendation systems, or data analytics must invest in unlearning capabilities to safeguard revenue models against privacy regulations while maintaining competitive AI performance.

c. Operational Scale Economics and Cost Structure Optimization

All six case studies showed substantial operational cost reductions—SBI saved ₹142 crores annually, Amazon achieved $47 million in savings, and Reliance Jio saved ₹347 crores. Business owners should view machine unlearning as essential infrastructure that enables AI scalability while controlling compliance costs. The technology allows organizations to process thousands of privacy requests without proportional increases in operational overhead, fundamentally changing the economics of AI deployment at scale.

6.4.2.2 For Tech Product Developers

a. Privacy-by-Design Architecture Requirements

The case studies reveal that retrofitting existing AI systems with unlearning capabilities requires significantly more resources than building privacy-preserving architectures from inception. Microsoft's healthcare platform and Jio's telecom infrastructure succeeded because they integrated unlearning capabilities into core system design. Tech developers must adopt privacy-by-design principles, incorporating unlearning APIs, data lineage tracking, and federated learning architectures as foundational components rather than aftermarket additions to AI products.

b. Domain-Specific Unlearning Frameworks Over Generic Solutions

Each successful implementation required specialized approaches—JPMorgan's generative AI focus, Amazon's federated commerce architecture, and SBI's agentic

AI memory management. Generic unlearning solutions proved inadequate for complex enterprise requirements. Tech developers should prioritize building domain-specific unlearning frameworks that address sector-unique challenges including regulatory requirements, data interdependencies, and performance preservation needs rather than pursuing one-size-fits-all approaches.

c. Real-Time Processing and Quality Assurance Integration

All case studies emphasized the critical importance of maintaining AI system performance during unlearning operations. Paytm's real-time transaction processing integration and Reliance Jio's network optimization preservation demonstrated that successful unlearning requires sophisticated quality assurance and performance monitoring capabilities. Developers must integrate continuous testing, automated validation, and rollback mechanisms to ensure unlearning operations don't degrade core AI functionality or user experience.

6.4.2.3 For Product/Solution Vendors

a. Enterprise Integration Complexity as Market Opportunity

The case studies highlight the extensive integration challenges organizations face when implementing machine unlearning across existing AI infrastructure. Amazon's SageMaker integration, Microsoft's Azure healthcare platform coordination, and SBI's hybrid cloud management represent significant technical complexity that creates substantial market opportunities for specialized vendors. Solution providers should focus on developing comprehensive integration platforms that can coordinate unlearning across diverse enterprise AI ecosystems rather than standalone point solutions.

b. Regulatory and Geographic Customization Requirements

Indian enterprise cases (SBI, Paytm, Reliance Jio) demonstrate the need for solutions that adapt to local regulatory frameworks, linguistic diversity, and market-specific requirements. SBI's multilingual data complexity and Paytm's state-level regulatory variations show that global vendors must develop localization capabilities extending beyond language translation to include regulatory compliance, cultural privacy expectations, and regional business practices. Vendors should build configurable compliance modules and regional adaptation frameworks to address diverse international markets.

c. Consultative Sales and Implementation Services Necessity

All six case studies required extensive organizational change management, stakeholder training, and cross-functional coordination beyond technical implementation. JPMorgan's compliance officer education, Microsoft's clinical validation protocols, and Reliance Jio's multi-business coordination demonstrate that successful deployments require comprehensive consulting services. Solution vendors must develop implementation methodologies, training programs, and ongoing support capabilities that address organizational transformation aspects of machine unlearning adoption, shifting from product sales to solution partnership models.

These implications indicate that machine unlearning represents a fundamental shift in AI architecture, business operations, and market dynamics rather than an incremental privacy feature. Organizations and vendors that recognize these strategic implications and adapt accordingly will capture significant competitive advantages in the emerging privacy-preserving AI economy.

Chapter 7
The Road Ahead: Ethical, Technical and Governance Perspectives for Holistic Machine Unlearning

Abstract Machine unlearning is becoming increasingly important as modern AI systems must keep up with strict privacy regulations, fast-changing datasets, and users who expect greater control over their information. This chapter brings together the major elements that make unlearning workable in real-world settings—ensuring it remains ethical, secure, scalable, and aligned with the expectations of responsible AI. On the ethical front, we examine issues of fairness, transparency, and user autonomy. Security adds another layer of complexity, because once models are allowed to "forget," the unlearning process itself can be targeted or exploited in new ways. From a systems perspective, unlearning also confronts practical challenges: large and complex models, decentralized setups like federated learning, and the need to build unlearning directly into ML frameworks rather than bolting it on as an afterthought. Governance frameworks then become essential, offering ways for organizations to document, verify, and legally demonstrate that unlearning has truly occurred. And since unlearning sits at the intersection of computer science, law, ethics, HCI, and public policy, interdisciplinary collaboration is key to making it trustworthy and reliable. Taken together, this final chapter ties up the major themes of the book—ethics, security, scalability, distributed learning, framework integration, interdisciplinary insights, and governance—to offer a clearer and more complete picture of what machine unlearning really involves. The overarching message is that unlearning is not a single algorithm or a one-step procedure; it is a coordinated effort that spans technical, organizational, and societal layers.

7.1 Introduction

The earlier chapters in this book have served to establish that as machine learning systems become increasingly popular, being embedded in everyday systems, and as they continue to learn from more personal and sensitive data, the ability to "unlearn" information has become just as important as learning itself. However, unlearning introduces challenges that extend far beyond mere algorithmic domains such as model retraining. For instance, a whole network of questions pop up that go beyond

R. Gupta et al., *Machine Unlearning: Concepts, Techniques and Applications*,
Studies in Computational Intelligence 1270,
https://doi.org/10.1007/978-3-032-18447-4_7

the model's code: Is the unlearned model ethically fair? Is it secure? Who checks whether the model has actually forgotten—completely forgotten? And what happens when unlearning needs to work across millions of devices?

This chapter builds on these questions and moves beyond the theoretical discussions explored earlier, shifting our focus towards what lies ahead for machine unlearning in practice. Here, the goal is to lay out the concrete requirements for making machine unlearning actually work in real-world settings. To do this, we bring together the final pieces—ethics, security, scalability, distributed learning, integrated framework design, interdisciplinary collaboration, and governance—to form a clearer understanding of the entire machine unlearning puzzle.

We structure the discussion around these seven major areas. We begin by looking at the ethical foundations of unlearning (Sect. 7.2), including concerns about user autonomy and how removing certain data points or individuals might affect fairness in the model. This naturally leads into security considerations (Sect. 7.3), since allowing models to "forget" creates new vulnerabilities and potential attack routes that must be anticipated and mitigated. From there, we turn to the technical realities: the scalability barriers that arise when unlearning must operate on large, complex models (Sect. 7.4), and the additional complications introduced in federated learning settings, where unlearning must happen in a decentralized and often unpredictable environment (Sect. 7.5).

The final sections of the chapter look at how unlearning can be embedded into practice. We discuss ways to integrate unlearning capabilities directly into ML frameworks (Sect. 7.6), making forgetting a built-in feature rather than an afterthought. This is followed by a look at interdisciplinary approaches (Sect. 7.7), since meaningful progress in unlearning depends on insights from law, ethics, human–computer interaction, systems engineering, mathematics and more. All of these threads ultimately tie into governance frameworks (Sect. 7.8), which help organizations verify, certify, and audit unlearning processes within legal and organizational boundaries. Taken together, these sections form a holistic roadmap for transforming machine unlearning from a mostly theoretical idea into a practical and reliable component of responsible AI.

7.2 Ethical Considerations in Machine Unlearning

7.2.1 Introduction

As machine unlearning matures, it is pertinent to weave ethics into the foundation of machine learning algorithms, rather than treating them as an afterthought, lest these algorithms become selective or careless, leading to results that are unfair and tinted with bias. In the subsections that follow, we introspect some important facets that demand ethically responsible machine unlearning (MUL) systems.

7.2.2 Transparency and Accountability

Ensuring transparency in the unlearning process implies designing MUL systems that allow users to understand how their data is being removed, what limitations the process and the methods of removal entail, and what consequences might follow. Achieving accountability largely means fixing the responsibility in case, there is a performance degradation in the model after unlearning or, there has been an introduction of risks to fairness and security. For maintaining internal accountability of organizations, and for external audits or scrutiny by the regulatory boards, it is pertinent that the organizations maintain documentation of the technical and ethical unlearning processes followed by them (Mercuri et al. 2022).

7.2.3 Trust and User Autonomy

Trust is the fundamental building block of an ethical relationship. The users must be able to trust that their data is completely erased and not just superficially removed or hidden. Transparency and Accountability, discussed earlier, serve as the foundation for establishing user trust in machine unlearning systems. For instance, confirmation of complete unlearning can work as feedback mechanism fostering transparency and user trust (Mercuri et al. 2022). The notion of user autonomy is integral for an ethical unlearning process. User autonomy dictates that data removal should be treated as a user's right rather than a mere concession. The users should be empowered to decide how and when their data is used or erased from a machine learning model.

7.2.4 Fairness and Bias

Unlearning can inadvertently amplify biases or create new disparities in machine learning systems, leading to skewed outcomes. As an example, let us consider a hitherto representative dataset, which after unlearning may become unrepresentative because the data erasure requests predominantly came from the same demographic group. Ethical unlearning demands ensuring fairness and avoiding bias in the unlearning processes. Fairness is particularly essential in specific sectors such as healthcare, recruitment, and lending. To ensure fairness of the unlearning process, it is important to carefully monitor demographic distributions before and after unlearning. Some corrective strategies such as reweighting or fairness-aware retraining may also be employed for restoring the fairness (Zhang et al. 2024, Oesterling et al. 2024).

7.2.5 *Privacy and the Right to Be Forgotten*

Regulations such as the General Data Protection Regulation (GDPR) and the California Consumer Privacy Act (CCPA) that enshrine the "right to be forgotten," stem from the ethical obligation of safeguarding user privacy. These compliances legally bind organizations to guarantee complete data erasure. If any subtle traces of the same are left, the organisations are set to face legal reprimand, disrepute and brand tarnishment. Hence, these laws go a long way in ensuring that the user's requests to erase their data are honored not just as a legal requirement, but as a moral obligation to respect individual autonomy and privacy.

7.2.6 *Security and Misuse Risks*

Adversarial attacks could be launched by nefarious individuals by misusing the deletion requests. These attacks are generally orchestrated as coordinated campaigns to delete data without consent from the owners in order to weaken or sabotage the MUL systems (Chen et al. 2021). Ethics demand that in order to safeguard user data from exploitation, such attacks be foreseen at the design time so that adequate security mechanisms can be woven into the design, leading to robust systems. For instance, provisions must be made to authenticate deletion requests before they are executed. This may include having an anomaly detection system to identify malicious patterns. Similarly, after large-scale unlearning, monitoring the model performance can provide clues to a malicious deletion attempt. The unlearned system can also be verified by designing feedback chains.

7.2.7 *Ethical Boundaries and Trade-Offs*

Machine unlearning is confronted with competing ethical values. A very challenging task is therefore to set clear ethical boundaries that can be used to interpret when unlearning is appropriate and when it may conflict with broader societal interests. For instance, while users need to have autonomy for exercising their 'right to be forgotten, the machine learning models built from these data objects are often deployed in public interest applications such as healthcare research or fraud detection. Users' call for data erasure leads to change in models, potentially affecting their efficiency or even reliability. If critical knowledge is erased, public safety or societal good may even be endangered (Zhang et al. 2023). Thus, it is challenging but pertinent to take an ethical stand when navigating the trade-offs between user rights and societal responsibilities.

7.2.8 Conclusions and Open Research Directions

To conclude, ethical considerations remind us that unlearning is not just a technical task but a commitment to fairness, transparency, and respecting the rights of the people behind the data. To facilitate the evolution of machine unlearning as a practice that respects individual rights while sustaining trust, fairness, privacy, and societal benefit, the following future directions promise fruitful results:

- In line with the principles such as privacy and fairness by design, the ethical obligations should be specified along with the technical requirements, in the design phase of the MUL systems (Oesterling et al. 2024).
- Subjecting MUL systems to independent audits and oversight mechanisms can play a vital role in ensuring accountability and fairness.
- Bringing users into a feedback loop can help catch unwanted or adversarial unlearning requests early, while also building greater trust by letting users stay involved in the process.
- Ensuring consistent application of unlearning principles and practices across organizations may help in better accountability.

7.3 Security Considerations and Mitigating Attacks in Machine Unlearning

7.3.1 Introduction

Security is a very important facet for machine unlearning systems, typically even more pronounced than for the traditional machine learning (ML) systems. In fact, it can be argued that if the deletion pipelines, request verification, and model's updates aren't properly secured, machine unlearning might end up undermining the very privacy and trust it was meant to protect in the first place (Chourasia and Shah 2023). While security in ML systems mainly focuses on risks across data collection, storage, training, and inference, MUL systems face a much wider and more complex attack surface. In machine unlearning systems, adversaries may attempt to exploit unlearning pipelines to infer private information, corrupt models, or disrupt availability. Sometimes, attackers selectively force the removal of influential examples. This causes distorted outcomes and undermines the unlearned model's integrity (Huang et al. 2024). In high-stake domains such as healthcare, finance, and law enforcement, these concerns are magnified because of severe ethical, legal, or financial harm implications. In the following subsections, we highlight the multiple dimensions of risk and then dedicate a section on how to mitigate these risks.

7.3.2 *Unlearning's Dark Side: Exploring Attack Surfaces and Security Threats in Machine Unlearning*

This section explores the types of key security risks and vulnerabilities for unlearning systems.

- **Denial-of-Service (DoS)-style Attacks**: In DOS attacks, a large number of unlearning requests are typically generated by malicious users to overwhelm system resources. Sometimes, the deletion requests may even be put up for fake data points. The main intent is to delay or halt legitimate unlearning operations and disrupt model availability.
- **Adversarial Unlearning Requests**: These requests are malicious deletion requests, crafted with the objective of harming the model rather than actually removing data (Zhao et al. 2024). For example, an attacker might submit unlearning requests for data that was never in the training set, thereby misleading the unlearning pipeline and degrading model accuracy substantially (Huang et al. 2024). In other scenarios, attackers might exploit weak request verification by requesting removal of many records from a specific subgroup, effectively introducing bias into the model. Some attacks aim to force the system to forget highly influential data points, causing the model's performance to collapse (Di et al. 2024).
- **Unauthorized or Impersonated Requests**: Illegitimate users gain unauthorized access to a genuine user's account and submit data deletion requests, usually to disrupt the system or remove data that is critical to the model's performance. This is called an impersonation attack. These unauthorised attacks may also be the cause of Denial of Service (DoS) attacks discussed before, where illegitimate users post requests to remove data that was never a part of the training set.
- **Model Extraction via Unlearning**: An attacker submits carefully crafted or repeated unlearning requests. By observing the changes in the model's parameters or behavior, upon serving the maliciously submitted requests, the attacker infers structure, weights and inner working of the underlying model and sensitive information about the training set.
- **Residual Information Leakage**: Incomplete or approximate unlearning methods, which are designed to be more computationally efficient than full retraining, do not fully remove the data's footprint and leave traces of supposedly deleted data. These traces, such as the inter-dependencies among records and cached computations, often allow for exploitation, as the deleted information can be reconstructed, thereby leading to undermining of user privacy (Chourasia and Shah 2023).

Let us consider a couple of illustrative case examples to make the above discussion on security risks understandable.

- Consider the case of incomplete unlearning in a healthcare dataset. A patient could be exposed to significant privacy risks if residual traces of sensitive information like a diagnosis, specific treatment plan or medical bills are not removed completely. A malicious actor could first use a membership inference attack to confirm whether patient's data was part of the training set, even after it was supposedly erased and then lead a data reconstruction attack whereby elements of the patient's record from the model's residual knowledge are pieced together, ironically defeating the very purpose of the unlearning request raised by a patient.

- As a second example, consider a credit risk assessment model that predicts a person's creditworthiness using their financial history and other personal information. Now imagine an adversarial deletion request coming in from a bot. A flood of such unauthorized requests could overwhelm the system and disrupt its functioning, essentially turning into a denial-of-service attack. If these deletion requests are targeted at specific demographic groups, the model could also become biased, causing its overall accuracy to drop (Huang et al. 2024). On the other hand, if an individual's request to unlearn their financial history is not fully carried out, the consequences can be equally serious. Incomplete unlearning may reveal traces of their personal data, creating the risk of sensitive information leakage. A reconstruction attack in this context would be extremely damaging, as it could expose confidential financial details and undermine user trust entirely.

7.3.3 Mitigation Strategies

The security risks and real-world scenarios discussed above show how even small lapses in unlearning can quickly turn into serious ethical and practical problems. In that sense, security becomes just as central to unlearning as privacy and fairness—without it, the entire purpose of forgetting breaks down. This makes it essential to think carefully about how to prevent, detect, and reduce the impact of these threats. This section outlines key risk-mitigation strategies that can help strengthen unlearning systems and make them more resilient in practice.

- **Robust Request Authentication**: This refers to the process of access control that verifies the identity and legitimacy of a user, so that only authorized individuals can ask for their data to be removed. The primary objective of robust request authentication is to prevent impersonation attacks. These attacks are automated requests designed to corrupt the model and sabotage the unlearning pipeline. Mechanisms for robust request authentication often rely on multi-factor checks to ensure that every unlearning request truly comes from a legitimate user. This can involve simple measures like password plus OTP or email verification, or stronger methods such as digital signatures that securely link the request to the actual data owner.
- **Rate Limiting**: Restriction limits complement access control by controlling the frequency and volume of unlearning requests that a user can submit within a given time frame. For example, limiting the rate of accessing APIs to prevent abuse through excessive or automated requests. Posing limits is a mechanism that prevents adversarial attempts that can overwhelm the unlearning pipelines, and therefore go a long way in preventing denial-of-service style attacks.

- **Cryptographic Protocols for Unlearning**: These are advanced security strategies that keep sensitive information secure by using encryption techniques. Secure Multi-Party Computation (MPC) technique ensures no one (not even the model owner) is able to directly know about any user's specific unlearning request. Homomorphic encryption ensures that a model can be unlearned without decrypting the data, ensuring the original information remains confidential throughout the entire process. Cryptographic protocols prevent residual information leakage, making it a powerful defense against a range of privacy and security attacks. An example cryptographic framework has been proposed by (Eisenhofer et al. 2022). This framework generates proofs of deletion using SNARKs and Merkle trees, providing verifiable guarantees that a data point has been completely deleted.
- **Robust Unlearning Algorithms**: These algorithms prioritize security from an algorithmic standpoint, complementing other mitigation strategies. Robust unlearning algorithms are designed to be resistant to privacy leakage and malicious unlearning attacks. Contrary to the mitigation approaches involving external cryptographic or system-level defenses, these algorithms incorporate security-enhancing properties directly into their design. For example, some robust algorithms might focus on eliminating residual information and providing certified unlearning by offering a guarantee that the unlearned model is statistically identical to a model retrained from scratch without the deleted data. Similarly, algorithms designed to be resilient to adversarial manipulation ensure that even if malicious requests are processed, the overall integrity and fairness of the model are preserved (Zhao et al. 2024).
- **Auditing and Logging**: These are crucial mechanisms for ensuring accountability and threat detection. A clear audit trail can be established by maintaining secure and tamper-proof logs of every unlearning request. This enables the real-time monitoring of all unlearning activity and post-incident analysis, allowing detection of suspicious or anomalous behavior, such as a single user submitting an unusually high number of requests or requests targeting sensitive, high-value data.
- **Penetration Testing (Pen-Testing)**: also known as Vulnerability Assessment or Regular Security Testing, is a proactive defense strategy for simulating real-world attacks. From simple integrity checks to sophisticated backdoor injections, these tests are designed to identify vulnerabilities before they can be exploited by malicious actors. The key difference from the other mitigation strategies is actively probing the system with a variety of attacks to test its resilience, which is unlike the passive logging or static algorithms discussed above. So, Pen-Testing helps to identify and fix failures before they lead to privacy breaches. An example is a probabilistic verification framework proposed by Sommer et al. (2020) that uses hypothesis testing to test for any residual malicious influence and employs backdoor attacks to see if an attacker can use them to manipulate the system.

7.3.4 Conclusions and Open Research Directions

Security in machine unlearning is as critical as privacy and fairness. The unlearning systems are susceptible to manipulations, disruptions, or theft of sensitive information from pipelines. By designing algorithms that use mechanisms such as robust authentication, encryption, rate limits and access controls etc., these threats can be mitigated. However, some open questions remain:

- There are no formal security guarantees for unlearning algorithms (Eisenhofer et al. 2022).
- Efficiency must be balanced with robustness since significant computational overheads are incurred to develop highly secure protocols.
- There is a lack of benchmark datasets and standardized frameworks for the evaluation of unlearned model's security (Sommer et al. 2020).

7.4 When Unlearning Meets Scale: Scalability Challenges in Machine Unlearning

7.4.1 The Core Problem

At the heart of machine unlearning is the aim of protecting user privacy by enabling removal of individual data points from trained machine learning models. But in practice, the very massive real-world datasets and intricate models pose a significant computational challenge to unlearn even a single record, sometimes requiring retraining the entire model from scratch. This makes it essential to design unlearning algorithms that scale gracefully in real-world settings. Scalability, is the ability of an unlearning algorithm to retain its effectiveness and efficiency with the increase in size of dataset, model complexity, and the number of deletion requests.

7.4.2 Unlearning Paradigms

One of the trade-offs in scaling machine unlearning frameworks is striking a balance between the two competing goals: completely removing data with mathematical precision (Exact Unlearning) and keeping the process computationally practical (Approximate Unlearning). This section explores how the contrasting nature of these two unlearning paradigms contributes to the practical difficulties of scaling machine unlearning.

- **Exact Unlearning**: This paradigm is widely recognized as the most trustworthy way to satisfy regulatory expectations, because it results in a model that has been fully retrained and completely purged of the deleted data. However, the computational cost of exact algorithms that perform full retraining grows at least linearly with the dataset size, making it economically unviable for large-scale, real-time systems.

- **Approximate Unlearning**: Approximate unlearning is far more efficient than exact unlearning, because it avoids full model retraining, using techniques like influence functions or selectively fine-tuning certain layers. This makes it much faster and computationally less expensive, but the trade-off is the unlearning deficit—small residual traces of the deleted data that may still influence the model and create security concerns.

7.4.3 Unlearning Under Scale: Challenges

This section deepens insight about the challenges that need to be addressed to aid the design of scalable unlearning algorithms.

- **Model Complexity**: A very big challenge for scaling machine learning is the complexity of the machine learning model it is trying to unlearn. A typical deep learning model such as a Large Language Model (LLM) has a very vast parameter space. There are millions or billions of parameters (weights and biases) that need to be identified, adjusted and verified after each unlearning request is completed. The computational time required to do this is O(N) or worse, where N is the total number of parameters in the model. As an illustrative example, let us consider a multilingual machine translation model based on a massive transformer architecture with billions of parameters, designed to translate 100 different languages. To honor a request to unlearn even a very tiny set of domain-specific English to German phrases, the unlearning algorithm must check, verify, and adjust the billion-scale parameter matrix.
- **Data Inter-dependencies**: In embeddings, graphs, or collaborative filterings, many observations are interconnected. As a result, data points are rarely independent and the influence of a single training data point is spread across the billions of parameters. A deep neural network (or large transformer/ graph model) with many layers, has non-linear and intertwined parameter dependencies. If a single record is removed, representations or similarity metrics change for many others, causing cascading model changes. Thus, data inter-dependencies make it computationally impossible to localize the effect of one data point. As an illustrative example, let us continue the example of the multilingual machine translation model. It is not possible to simply unlearn the English-to-German connections because of the high complexity of the model, as mentioned before, and the interdependence of all languages within the same neural network structure. Moreover, after the unlearning of required phrases, a massive, computationally expensive verification has to be run across the entire model to ensure that adjusting the English-to-German weights didn't accidentally degrade the quality of other language translations. Latest research efforts are geared towards resolving the

unscalablity problem. For instance, (Chen et al. 2022) introduce planned isolation into the otherwise highly entangled structure of a Graph Neural Network (GNN). It mitigates the scalability problem by using approximate partitioning (sharding). (Wu et al. 2023a) have proposed Graph Influence Functions (GIF), that mathematically determine how the removal of one data point would affect the parameters of all other connected nodes in a GNN. This avoids the necessity of checking and re-adjusting the vast parameters.

- **Dataset Size**: The dataset sizes extend up to several petabytes in applications such as e-commerce, social media platforms, and large AI models. Often cited as the Big Data Dilemma, the traditional full or partial retraining used for deletion of a data point becomes infeasible, largely due to two factors- the sheer volume of data itself and the data inter-dependencies embedded in model. Removing a data point without tracking and removing all its correlations and effects on the underlying model leads to residual information leakage. While managing the concerns of big data and information leakage, it is also important to maintain the model's performance on the remaining data. Let us consider a couple of example scenarios to understand the above concerns. The computational overhead required to honor a deletion request of the X-ray or MRI of one of the patients from a hospital-wide diagnostic model, trained on terabytes of medical imaging data, would be equivalent to retraining the model from scratch. Similarly, an unlearning request to remove the browsing history of one of the billion users a recommendation engine is trained on, will be computationally prohibitive. So some approaches that avoid retraining are being studied. For instance Li et al. (2023) propose a method based on influence functions for recommendation unlearning that attempts to avoid retraining by estimating the impact of the data point to be deleted.
- **Frequency of Unlearning Requests**: When unlearning is treated as a continuous service, the high frequency of unlearning requests turns machine unlearning into a constant, high-volume data management problem. This represents a major operational and scalability challenge, because adjusting and verifying the model after every unlearning request takes time, during which the computational resources are locked. The system gets perpetually stuck if the time needed to process the unlearning operation exceeds the arrival rate of new requests. The model, therefore, may not even be longer able to provide conformance and the effectiveness guaranties, let alone scale up to the efficiency standards. To solve the high frequency problem, approaches like SISA (Sharded, Isolated, Sliced, Aggregated training) (Bourtoule et al. 2021a) partition the training data into isolated shards that handle unlearning requests locally, avoiding full model recalculation and providing significant speedup. However, the efficiency gained in unlearning speed may be offset by the increased overhead of managing, updating, and coordinating numerous shards, each potentially requiring its own checkpoints, version control, and security logs.

7.4.4 Mitigation Strategies for Scalability

In this section, we discuss some methods designed to reduce the severity of the impact of the scalability challenges discussed in the previous section, by reducing their computational burden, lowering the chances of privacy leakage and managing their operational complexity.

- **Algorithmic Innovations**: These innovations focus on reducing the computational scope of each unlearning request. Techniques like Sharding or SISA (Sharded, Isolated, Sliced, Aggregated training) work by dividing data into shards to localize unlearning, that reduces the computational cost per unlearning request substantially. Using influence-function based techniques allows complex structures like recommendation systems and graph models to offer substantial efficiency gains by calculating the minimal changes needed to specific model parameters without resorting to full retraining.
- **Architectural and System-Level Approaches**: These strategies focus on shaping the system's operational environment so it can handle continuous unlearning requests more efficiently. Model partitioning or modular design plays an important role here, because it breaks a large model into smaller, more independent components. This way, unlearning something in one part—say, a language-specific encoder—doesn't force unnecessary updates across the entire model. Alongside this, on-demand unlearning services can help manage heavy workloads by batching or prioritizing deletion requests. Instead of processing every request the moment it arrives, the system can group multiple requests together and run a single, more efficient update during off-peak hours. This approach lowers the computational cost and keeps the unlearning pipeline responsive even when requests are frequent.
- **Data-Centric Approaches**: This category of methods focuses on preparing the data in advance so that unlearning becomes easier, faster, and less resource-intensive later on. The basic idea is simple: if the model doesn't store too much raw or sensitive information during training, then removing a user's data later will be far less complicated. One way to do this is by training models on summarized or compact representations of the data. This means that when an unlearning request arrives, the system only needs to adjust a smaller set of encoded features instead of reprocessing large amounts of raw data. In addition to this, training the model with privacy-preserving or unlearning-friendly techniques—such as Differential Privacy or Federated Learning—helps control how much precise, user-specific information is encoded in the model. Techniques like Differential Privacy deliberately add noise during training, so the model relies less on exact individual data points and more on general patterns. Federated Learning, on the other hand, keeps raw data on the user's device and only shares aggregated updates, which naturally limits the amount of fine-grained personal detail that reaches the central server. Because these methods reduce the depth and sensitivity of what the model "remembers" about any one user, they also reduce the unlearning deficit i.e., the leftover influence that can linger even after a deletion request. In simple terms,

the less tightly a model is tied to individual records during training, the easier it becomes to remove those records later without leaving traces behind.

7.4.5 Conclusions and Open Research Directions

Scalability remains one of the biggest obstacles to deploying machine unlearning in real-world systems. The examples we have seen—from recommender models to graph neural networks and large-scale deep learning—show that techniques like influence functions, SISA training, and graph-based partitioning offer meaningful progress. However, each handles only a part of the problem. Modern ML systems are too large, too interconnected, and too dynamic for any single approach to handle all aspects of scalability on its own. In practice, scalable unlearning will demand a blend of algorithmic improvements, system-level engineering, and smarter data management strategies. Moving forward, the research community needs to focus on approaches that balance unlearning accuracy, privacy guarantees, and computational feasibility. Several promising research directions stand out:

- Sub-linear scalability: Developing algorithms whose unlearning cost grows slower than the dataset size or the number of deletion requests.
- Architecture-aware unlearning: Designing unlearning methods that are tailored to the structure of modern model architectures—such as transformers, large language models, graph neural networks, and recommender systems—is another important direction. Each of these models has its own internal wiring, dependencies, and ways of storing information. For instance, transformers and LLMs distribute knowledge across many layers and attention heads, making it hard to isolate the influence of a single data point. GNNs, on the other hand, propagate information across connected nodes, so removing one entry can unintentionally affect many others in the graph. Recommender systems embed users and items in shared spaces, meaning one user's deletion may shift an entire neighborhood of embeddings. Because of these differences, a one-size-fits-all unlearning method is unlikely to work. Instead, the field needs techniques that understand how each architecture stores and spreads information so that unlearning can target only the relevant components without disrupting the whole model.
- Handling high-frequency requests: Building low-latency pipelines that can manage continuous or streaming deletion requests without overwhelming the system.
- Modeling data interdependencies more accurately: Especially in graphs, collaborative filtering, or embedding spaces, where removing one point can ripple across many others.
- Better data partitioning and model modularity: Structuring models so that unlearning affects only local components, rather than triggering expensive global changes.

- Improved approximation guarantees: Reducing the "unlearning deficit" by offering clearer bounds on residual influence left behind after approximate updates.
- Comprehensive benchmarks and standardized datasets: Establishing shared evaluation suites to test scalability trade-offs—accuracy, speed, memory, and privacy—across diverse domains.

7.5 Unlearning in Federated Learning Environments

7.5.1 Introduction

Federated Learning (FL) is a decentralized paradigm that involves the training of a collaborative machine learning model across the central server and multiple cutting edge devices that act as clients, such as smartphones, medical sensors and institutional servers. Each client trains a local machine learning model on the data available locally with them. A collaborative model is then trained across the server and clients, with each client sending the important gradients of their local model or the updates thereof to a central server for aggregation.

It is important to note that this procedure of collaborative model training does not involve any direct raw data transfer or sharing. Resultantly, federated unlearning provides strong privacy guarantees, reduces data transfer overhead, making it increasingly popular in sensitive domains such as healthcare, finance, and mobile applications that require strict data protection. On the other hand unlearning also creates a fundamental conflict within the federated learning paradigm: data localization, allegedly one of the primary strengths of federated learning turns into it's greatest weakness while unlearning. Also, verifiable unlearning guarantees cannot be provided because the server cannot directly trace or reverse the contribution of a specific client or data instance, since it never accesses raw data and only maintains aggregated updates. Let us understand in the subsections that follow, the basic concepts underlying federated unlearning, how unlearning takes place in a federated system, what are the key challenges faced therein and what are some promising future directions of work in this field.

7.5.2 Foundations of Federated Unlearning

7.5.2.1 Motivation for Federated Unlearning

The motivation for federated unlearning stems from both ethical and legal realms. In accordance with the requirements laid down by regulations such as the General Data Protection Regulation and the Right to be Forgotten the users must be guaranteed the complete removal of their data from the federated nodes when requested.

7.5.2.2 Objectives of Federated Unlearning

The primary goal of federated unlearning is to ensure complete removal of requested user data from the local nodes and the removal of parameter traces that are stored on the central server, in the form of aggregated updates. The secondary goal is to minimize the retraining cost of the unlearned model and the communication overheads involved in the process. The tertiary goal is to help preserve privacy and fairness among remaining participants by preventing the introduction of bias or imbalance in the global model after selective unlearning.

7.5.2.3 Federated Versus Centralized Unlearning

In centralized learning, all the data is collected and processed in one place, while in federated learning, the data is distributed across multiple client devices, and only model updates are shared with the central server. The unlearning process is therefore straightforward in centralized learning systems—data can simply be deleted upon request, and the model retrained from the remaining dataset. In contrast, for federated unlearning, the separation between local and global parameters makes it difficult to identify and reverse the specific contribution of any one client or data point.

The central unlearning process can be seen as a single step process, whereas, federated unlearning process can be broken into two stages. The first stage—local deletion, i.e., the deletion of the data from the client (local device) is easy but requires care. The deletion of highly sensitive data from participating devices must be guarded against the possible leftover of the residual traces that can compromise privacy and the regulatory compliances. Second, and the more complex stage is the global unlearning that involves removing the influence of data from the global, aggregated model residing on the central server. This task is further complicated by communication bottlenecks and the limited visibility the central server has into individual nodes with client data. Also, the updates and aggregation in federated systems are asynchronous, since clients may train and communicate with the server at different times. Such uneven participation makes it very difficult to determine precisely when or how a client's data has influenced the global model. All these factors go into making federated unlearning a multi-layered and technically demanding process.

7.5.3 How Federated Systems 'Forget': Key Approaches

As discussed in the previous section, removing a user's data is difficult in federated learning environments. The different strategies offer a tradeoff in practicality and efficiency and can be broadly categorized into three categories—retraining-based approaches, approximation-based approaches, and hybrid approaches.

7.5.3.1 Retraining-Based Methods

Full retraining based methods rebuild the model from scratch, using all the data except the data to be deleted. This method ensures complete data removal but is extremely slow and resource-intensive, especially in large networks. Partial retraining also tries to rebuild the model but only specific layers of the model at a time or only the affected clients instead of redoing the entire process. This selective retraining reduces computation but may not always remove all traces of the old data perfectly.

7.5.3.2 Approximation-Based Methods

Approximation-Based Methods use mathematical functions to estimate and remove the influence of a client's data instead of retraining the model. Popular approximations use gradient-based or influence-function updates to "subtract" a user's contribution from the global model. Others adjust the weights or parameters to rebalance the model without a full retraining cycle. These techniques are definitely faster and use lighter computations, but they usually trade off a bit of precision for speed.

7.5.3.3 Hybrid (Client–Server) Unlearning Approaches

Most current unlearning methods stick to either the client side unlearning or the server side unlearning because coordinating both is too complex and costly. In server-side unlearning, the central server modifies the global model directly, often without involving the clients. This makes it faster and easier to manage, though less precise. In contrast, client-side unlearning asks each participating device to locally adjust or retrain its part of the model, which is more accurate but much heavier in terms of communication and bandwidth. However, as discussed in the previous section, ideally, federated unlearning should take place in two stages, i.e., both the affected client and the server should update the model at their end. Researchers are experimenting with hybrid methods that blend the two. For instance, efficient update compression and better aggregation schemes are being researched to shrink the size of the messages sent between the clients and the server, aiming to strike a balance between efficiency, accuracy, and privacy by letting the server and clients share the work intelligently.

7.5.4 Key Challenges and Considerations Before Federated Unlearning Systems

7.5.4.1 Ensuring Privacy

It is essential to ensure that the process of federated unlearning, i.e., the very process responsible for removing user data to protect their data does not accidentally leave any clues about the removed data. For instance, when clients adjust their models to forget data, the updates they send to the server could inadvertently expose hidden details about the data they kept.

7.5.4.2 Verification Guarantees

Since there is no explicit data sharing in federated learning and the central server cannot inspect the client's local data or local model state for residual traces, it has no direct way of ensuring data deletion. The entire unlearning process is based upon the trust that every decentralized device performed the deletion perfectly and completely. On the downside, a single malicious client could sabotage the entire unlearning process.

7.5.4.3 Communication Efficiency

In large federated learning networks with thousands of devices, communication poses large overheads. For every unlearning request, a model update is triggered and the server coordinates with all affected clients, which can take a lot of time and bandwidth—especially when connections aren't stable. To address these issues, researchers are exploring ways to make updates lighter, like compressing model changes and improving aggregation methods, so synchronization across the network becomes faster and more practical.

7.5.4.4 Asynchronous Client Participation and Drift in Client Updates

A big challenge in federated unlearning is that client devices, like phones or IoT gadgets, aren't always online. If a client that needs to take part in the unlearning process goes offline or drops out, the whole update can get delayed or stuck. This makes it hard for the server to complete or verify that the model has truly "forgotten" the data.

7.5.4.5 Non-IID and Heterogeneous Data

Data across clients is often heterogeneous and non-IID (non–independent and identically distributed). This implies that each client's data can be very different, and therefore its removal may lead to removal of a group with unique patterns. This may offset the balance of the underlying ML model by introducing bias and degrading performance for other, similar clients. As a result, it is tricky to make the model truly forget certain data while still keeping it accurate and fair for everyone else.

7.5.4.6 Security Risks

As in any other form of unlearning, security is a major concern in federated unlearning. For example, someone could send fake unlearning requests, pretending to be a real user — to mess with or weaken the global model. An adversarial attack can be launched by sending multiple fake requests to overwhelm the resources of a system. Membership inference attacks could exploit any clues left during the small tweaks made during unlearning. These help the attackers guess which data was kept in the model, even after the data was deleted.

7.5.5 Conclusions and Open Research Directions

Though federated unlearning is still in its infancy, when mature, the field could form the backbone of a more ethical and trustworthy era of decentralized AI. The open research directions discussed below highlight how most current solutions can be upgraded to balance privacy, accuracy, and scalability. The real progress will likely be visible by blending ideas across machine learning, distributed systems, and cryptography to create models that can truly "forget" responsibly—without slowing down or compromising fairness.

- **Generating Verifiable Proof of Forgetting**: One of the promising research directions for federated unlearning is developing formal ways to verify that the unlearning has actually worked. As discussed in the previous section, current methods often rely on trust rather than proof. Future research could explore mathematical or cryptographic guarantees that confirm a user's influence is gone, without retraining the entire model or exposing private data.
- **Efficiency of Scaling Federated Unlearning**: Scaling unlearning to real-world federated networks remains a huge challenge. Researchers are exploring how to make unlearning algorithms faster and lighter, possibly by reusing intermediate updates or compressing model adjustments, so that millions of devices can participate efficiently.
- **Auto correcting Bias**: Removal of heterogeneous data from the clients, risks drifting of underlying models towards bias or uneven performance. A worthy

research direction is creating adaptive mechanisms that monitor and correct such shifts automatically, ensuring fairness across the remaining data.

- **Adaptive Aggregation Mechanisms**: Another emerging idea is to build adaptive aggregation algorithms. When an unlearning request is received by a federated learning server, adaptive aggregation mechanisms subtract or rebalance the influence of the parameters on the server, depending upon the data to be deleted from the affected clients. This is in contrast to static aggregation where the global model undergoes very expensive retraining from scratch. An ideal adaptive aggregation mechanism design would save time (no full retraining), maintain stability (not mess up the model's performance), and preserve privacy (no need to share or re-access client data).
- **Standardization and Evaluation Frameworks**: Finally, consistent ways to test and compare unlearning methods are required. Tracking of progress and reproducible research can be facilitated by creating public benchmarks, open datasets, and clear evaluation metrics.

7.6 Integrating Unlearning Capabilities in Machine Learning Frameworks

7.6.1 Introduction

The preceding chapters in this book make it abundantly clear that as machine learning systems mature, unlearning—whether due to privacy regulations, user-initiated data withdrawal, or correcting flawed training data—has become just as essential as learning itself. Therefore, integrating unlearning into Machine Learning (ML) frameworks cannot come as an afterthought. Ideally, the unlearning components should be built directly into the ML frameworks so that developers are not forced to patch in custom, complicated code every time a deletion request comes in. Such ad-hoc unlearning solutions are slow, error-prone, and rarely scale well.

Yet, the current landscape is far from ideal. For instance, most popular frameworks like TensorFlow and PyTorch were built with learning in mind—not forgetting. Adding unlearning capabilities often requires rethinking parts of the model architecture, data-tracking mechanisms, and training pipelines (Bourtoule et al. 2021a). Having said this, let us take up the detailed discussion of the changes required to make unlearning a natural part of ML frameworks in the following subsections.

7.6.2 *Architectural Considerations*

7.6.2.1 Modular Design

As asserted in the introduction, Machine unlearning has to be treated as a core feature, not an add-on, and built into the ML architecture from day one. Breaking a model into smaller, independent pieces, i.e., modules, layers, or data shards, makes it easier to isolate and remove the effect of particular data. When a data point is requested for deletion, only its local shard needs to be retrained, not the entire model. Thus, a modular setup supports selective and scalable unlearning without rebuilding the entire system from scratch. For example, the SISA (Sharded, Isolated, Sliced, Aggregated) method provides a framework that divides data into shards and slices so that unlearning is more efficient by focusing only on affected parts (Bourtoule et al. 2021a).

7.6.2.2 Data Traceability

Tracking which data points influenced which parts of the model helps when one needs to remove them or their effects. The unlearning algorithm can freeze large portions of unaffected parts during the update, saving massive amounts of computation and memory. Metadata tagging and influence mapping are two techniques that can be used during training to make targeted data removal more efficient and verifiable (Zhang et al. 2024). Metadata Tagging involves attaching a unique identifier (a tag) to every data point during training, allowing the system to instantly know which model update came from which specific piece of data. Influence Mapping involves mathematically calculating the location and amount of a single data point's impact across the model's entire set of parameters.

7.6.2.3 Differential Privacy Integration During Unlearning

Differential Privacy is a technique that provides a provable, measurable level of protection against sophisticated attacks (like membership inference). It involves adding a small, calculated amount of deliberate noise to the data, so that a mathematical guarantee can be provided that the unlearned model won't reveal whether any specific person's data was included in the calculation. The idea is that Differential Privacy shouldn't stop after training—the framework should also use it when performing unlearning updates. This extra layer of security helps mask any residual influence and ensures that the unlearning update itself doesn't expose sensitive details, giving a verifiable privacy guarantee.

7.6.3 *Framework-Level Modifications*

The ML frameworks need specific features to support unlearning efficiently.

7.6.3.1 Standardized Built-In Unlearning APIs

Future framework versions could include simple, non-expert functions to trigger deletion. For instance, native methods like unlearn() akin to fit() or predict(), can be included, that abstract the complex unlearning algorithm behind a single, consistent call. This would make unlearning far easier for everyday developers and teams, encouraging wider, mainstream use of unlearning features across different applications. In other words, such simplicity is what enables true mass adoption—because when something is easy to use, more people actually use it, and they are far less likely to make mistakes.

7.6.3.2 Efficient Checkpointing and Version Control

Keeping a history of model states (checkpoints) allows quick rollbacks rather than full retraining. Using differential checkpoints, i.e., storing only incremental changes such as the necessary data partitions (shards) or influence summaries required for fast partial retraining (like SISA), reduces storage and computation overhead.

7.6.4 *Automation and Tools*

7.6.4.1 Automated Data Lineage Tracking

Automated tools that log data origin, usage, and it's influence on the model should be integrated with the ML framework so that unlearning requests can be handled with less manual effort and better accuracy.

7.6.4.2 Integration with Privacy and Compliance Modules

The deletion workflows should be automated by linking ML pipelines with regulatory frameworks (e.g., GDPR or RTBF), so that, when data has to be removed, the ML system is already set up to handle it.

7.6.5 *Performance and Scalability Trade-Offs*

7.6.5.1 Balancing Speed and Accuracy

Unlearning operations should not come at the cost of significant drops in model accuracy. Approximate methods can make the process much faster, but they sometimes sacrifice precision, so finding the right balance between speed and accuracy becomes crucial (Dilworth 2025).

7.6.5.2 Resource Optimization

To handle a large number of unlearning requests, frameworks can use strategies like lazy retraining, caching, and distributed updates. Lazy retraining is an optimization technique where the system defers the actual unlearning computation until multiple requests have accumulated, thus processing deletion requests in batches to save computational and communication resources. Caching (temporary storage of intermediate computations) and distributed updates (non-centralized processing) are system-level optimizations, used to significantly speed up the iterative steps required during the unlearning process.

7.6.6 *Evaluation and Verification*

7.6.6.1 Defining Metrics

Standardized metrics—such as residual influence, accuracy drop on retained data, or leakage of private information—need to be developed to measure if an unlearning approach really works, and whether it can be compared across systems.

7.6.6.2 Transparent Reporting

Frameworks should produce logs or certificates showing that specific data has been removed and it's model influence mitigated. This kind of transparency builds trust and helps with audits and compliance.

7.6.7 *Conclusions and Open Research Directions*

Adding unlearning to mainstream ML frameworks isn't just a technical upgrade, it's a mindset shift. The path ahead lies in designing frameworks that support learning and forgetting equally. Future ML systems can be geared for graceful forgetting by weaving in the features such as modularity, traceability, automation, and compliance integration discussed above.

Open Research Directions

- **Better data tracking inside frameworks**: Building simple, automatic ways for frameworks to keep track of how each data point influences different parts of the model, so that unlearning knows exactly what to update.
- **Unlearning-aware training pipelines**: Rethinking data loading, batching, and checkpointing, so that these steps are more friendly to future deletion requests instead of being optimized only for speed.
- **Rollback mechanisms**: Exploring model components or layers that can be reversed or updated independently, without touching the whole network.
- **General-purpose unlearning APIs**: Creating basic, framework-level functions that work across models—so that something like unlearn() feels as natural as fit().
- **Detecting unlearning-induced drift**: Adding simple monitoring tools inside frameworks to flag when too many deletions start shifting the model away from its intended behavior.

7.7 Interdisciplinary Approaches to Advancing Machine Unlearning Research

7.7.1 Introduction

Machine unlearning sits at the intersection of technology, law, and ethics, and it needs solutions that can satisfy all three—practical code, regulatory expectations, and the broader ethical principles behind data rights (Serrano and Li 2022). So, to build responsible unlearning systems, researchers need both technical innovation and social awareness (Xu et al. 2023a, Aslam et al. 2025). While core computer science is required to build the algorithms, mathematics helps prove their feasibility, and law is required to mandate exactly what will work legally. A technical fix that can't hold up in a legal review is basically useless—and a law that can't be put into practice through real algorithms, is just an idea on paper. Real progress in unlearning will only happen with interdisciplinarity, i.e., when technology, law, and ethics come together to build systems that are efficient, provable, and meet legal standards. In this section, we explore how the very different disciplines synergise unlearning and can contribute to advancing the field in a holistic and responsible way.

7.7.2 *Legal and Regulatory Perspectives*

7.7.2.1 Data Protection and the Right to Be Forgotten

The primary need for verifiable and efficient unlearning stems from the privacy laws and regulations such as the General Data Protection Regulation (GDPR) and the California Consumer Privacy Act (CCPA), that empower individuals to request data deletion in the first place (Xu et al. 2023a). Hence, privacy laws and regulations serve as a major catalyst for machine unlearning.

7.7.2.2 Compliance and Accountability in Practice

The legal and technical communities must collaborate to develop standardized compliance protocols and reporting mechanisms to provide proof of unlearning. For example, keeping clear audit trails helps show that unlearning is carried out properly, stays in line with data-protection laws, and can be reliably verified by privacy auditors (Barthwal et al. 2025).

7.7.3 *Ethical and Philosophical Foundations*

7.7.3.1 Balancing Individual Rights and Collective Benefit

As discussed in Sect. 7.2.7, unlearning raises ethical trade-offs. For instance, it must be deliberated that should an individual's request for data removal always outweigh the collective benefits of models, especially used in sensitive domains such as healthcare and safety. Ethics helps to negotiate these boundaries, ensuring that unlearning respects personal rights without undermining social good (Zhang et al. 2024, Aslam et al. 2025).

7.7.3.2 Fairness, Transparency, and Moral Responsibility

Ethics requires researchers to consider how to unlearn without affecting fairness or introducing bias. As discussed in Sect. 7.2.4, removing data might unintentionally remove the data of a particular sect, cast or creed, thereby skewing model outcomes and reducing representativeness. Therefore, philosophical perspectives on responsibility and fairness must be integrated into the design of machine unlearning systems that can guide the design of more transparent and equitable systems (Zhang et al. 2024).

7.7.4 Computer Science and Engineering

7.7.4.1 Algorithmic Efficiency and Optimization

The basic idea behind algorithmic optimization in unlearning is to make the process faster and less costly than retraining the entire model from scratch. This can be achieved by designing efficient approximate unlearning algorithms (Bourtoule et al. 2021b) or by using techniques like the lazy retraining and caching, that defer and batch deletion requests, saving massive computational costs.

7.7.4.2 System Architecture Integration

As discussed in Sect. 7.6.3.1, the objective behind system-level architecture modifications should be to integrate unlearning directly into the tools developers use. For instance, building in the native unlearning APIs (model.forgetUser()) in the development frameworks like the PyTorch and TensorFlow. Other, system-level architectural optimizations like sharding and distributed updates help to isolate and localize unlearning requests, saving the massive cost of retraining from the scratch (Bourtoule et al. 2021b).

7.7.4.3 Human–Computer Interaction (HCI) and User Experience

It should be easy for a user to request unlearning of a data point and verify it. HCI research supports creating easy and unambiguous interfaces for deletion requests, notifications, and progress tracking, thereby building trust in AI systems (Barthwal et al. 2025). Ideally, HCI should also communicate model limitations to users—for instance, clarifying that while data has been removed, residual effects might still exist. Making such transparent communication helps set realistic expectations (Aslam et al. 2025).

7.7.5 The Mathematical and Theoretical Foundations

7.7.5.1 Verifiable Guarantees and Proofs

Mathematical soundness provides the necessary rigor and verifiable proof that the data has been truly erased. For instance, metrics and tests have been developed to show that a model is statistically indistinguishable from the one never trained on the deleted data. Complexity theory is being used to formally define the boundaries of exact unlearning (Neel et al. 2022).

7.7.5.2 Information Theory and Quantifying Influence

Mathematics also helps to measure what precisely needs to be removed. Using ideas from information theory, the residual influence of a data point on the model's weights can be calculated. Graph Influence Functions (GIF) ensure that we only need to adjust the parameters that were actually affected, thereby avoiding wasteful global checks (Wu et al. 2023b).

7.7.6 Social and Behavioral Sciences

7.7.6.1 Understanding User Perception of Privacy and Trust

Insights from psychology and sociology can guide the design of user-friendly and trustworthy unlearning interfaces (Barthwal et al. 2025). For instance, behavioral science helps to understand how users perceive data deletion, privacy, and trust.

7.7.6.2 Public Policy and Societal Implications

Unlearning also has broader social consequences. Removing data at scale can influence public policy decisions, affect long-term research datasets, or even reshape how communities preserve information over time. Because of this, social scientists and policymakers need to work closely with technologists to create governance frameworks that are fair, sustainable, and aligned with societal expectations (Aslam et al. 2025).

7.7.7 Conclusions and Open Research Directions

To conclude, the larger message running through the discussion in this section is that real progress in unlearning will only come from genuine interdisciplinary collaboration. Technical tools alone cannot carry the full burden. Data scientists, ethicists, legal scholars, behavioral researchers, and system designers all contribute different parts of the solution, and unlearning becomes far more practical when these perspectives work together. This ensures that unlearning is not an afterthought but something built directly into the ML lifecycle in a way that respects user rights, regulatory rules, and broader societal expectations (Xu et al. 2023a).

At the same time, the field needs stronger standardization. Common benchmarks, open datasets, and transparent evaluation protocols would allow researchers to compare methods more fairly and understand the trade-offs between accuracy, speed, privacy, and fairness (Xue et al. 2025). Without shared standards, many solutions risk remaining isolated or difficult to reproduce.

Open Research Directions

- **Shared Benchmarks and Evaluation Suites**: Developing standardized datasets and metrics to compare unlearning performance across domains.
- **Interdisciplinary Governance Models**: Creating frameworks that combine legal, ethical, and technical constraints into clear operational guidelines for unlearning.
- **Human–Centered Unlearning Interfaces**: Designing user interactions and consent workflows informed by behavioral science and psychology.
- **Privacy-Preserving Verification**: Building methods that let auditors confirm unlearning without exposing private information.
- **Societal Impact Modeling**: Studying how large-scale unlearning affects public datasets, research integrity, and long-term societal memory.
- **Cross-Domain Collaboration**: Encouraging research that integrates computer science with law, ethics, sociology, and public policy to address unlearning holistically.

7.8 Governance Frameworks for Machine Unlearning

7.8.1 Introduction

The rapidly evolving landscape of data protection and privacy-preserving AI has made machine unlearning an essential and critical capability. However, realization of unlearning is not possible through technical methods alone. Strong governmental frameworks are needed to mandate procedures and also provide a robust backbone for unlearning systems. These frameworks outline specific responsibilities, verification methods, and compliance pathways, ensuring that unlearning is conducted accurately, documented properly, and can be audited (Bourtoule et al. 2021a, Xu et al. 2023b). With the help of these frameworks, high-level privacy principles translate into everyday operational rules that organizations can follow reliably (Voigt and von dem Bussche, 2017). Without these clear rules, organizations risk both non-compliance fines and a loss of public trust.

7.8.2 Legal and Regulatory Governance

7.8.2.1 Legal Mandates for Unlearning

Addressing the legal mandate is the primary motivation for establishing a governance framework. Regulations such as GDPR and CCPA require organizations to support data deletion and define how unlearning is verified (Carvalho and Ribeiro 2023). Verification mechanisms include influence-based tests, reproducible deletion logs, and cryptographic proofs that show a user's data no longer influences the model

(Gupta and Nguyen 2021). The proof of deletion is often provided through a Certificate of Unlearning (Patel and Chen 2025). The governmental frameworks must also define the validity and scope of an unlearning request (Garcia and Lee 2023). The framework should also establish Service Level Agreements for the time taken to complete the unlearning process and report back to the user.

7.8.2.2 Accountability and Enforcement

Governance frameworks should define who is accountable when unlearning fails. This typically involves appointing a Data Protection Officer (an individual who monitors and advises on compliance with data protection laws such as GDPR), the Machine Learning Engineering team, and legal counsel (Zhu and O'Connell 2024). Additionally, Data controller (the business or organization that holds the primary legal responsibility for law compliance), AI oversight boards, and external auditors, all play a role in ensuring that deletion requests are fulfilled in a verifiable manner (Tjoa and Guan 2020).

7.8.3 Organizational Governance Mechanisms

7.8.3.1 Request Intake and Validation Workflow

A well-defined process is needed to receive, authenticate, and prioritize unlearning requests, distinguishing them from simple data deletion in the raw dataset.

7.8.3.2 Internal Oversight Structures

The regulatory requirements must be converted into actionable workflows to be effective. Inside the organizations, this job is performed by forming internal review boards, privacy teams, and risk committees. These groups manage unlearning requests and track their completion.

7.8.3.3 Audit Trails

Strong governance mandates the Audit Trails, i.e., automatic maintenance of logs regarding unlearning events, including what model updates were made, when they were made, and how data was removed (Shokri and Shmatikov 2015). These records are crucial for satisfying regulatory bodies during an audit or investigation into compliance. Audit trails also help ensure transparency and consistency across large teams.

7.8.3.4 Secure and Transparent Infrastructure

The challenge is to design verification tools that are trustworthy without being too heavy or expensive to run. Blockchain-based audit trails or secure logging mechanisms are generally designed to create tamper-proof evidence of unlearning (Rao and Patel 2023).

7.8.4 Ethical and Cross-Disciplinary Governance

7.8.4.1 Fairness and Social Impact

Ethical oversight helps ensure that unlearning protects individuals without harming model equity. For instance, ethical governance ensures that the unlearning does not create any new bias or fairness issues in case certain marginalized groups frequently request deletion (Zhang et al. 2024).

7.8.4.2 Interdisciplinarity and Collaboration

Interdisciplinary review ensures that unlearning methods remain aligned with both societal expectations and technical feasibility (Brundage 2020). So, true governance requires cooperation among legal experts, ethicists, mathematicians, and machine learning researchers.

7.8.5 Conclusions and Open Research Directions

Overall, the future of unlearning depends as much on governance as on algorithms. Strong governance frameworks bring together policy guidance, organizational processes, technical verification, and ethical oversight. When combined, these elements help ensure that unlearning is not only technically feasible, but also reliable, transparent, and aligned with social expectations and regulatory obligations. There is still no universally agreed-upon definition of what "sufficient unlearning" actually means. Existing verification techniques remain fragile, and many organizations simply do not have the infrastructure to support consistent auditing or reliable proof of compliance (Cao et al. 2024). This gap makes governance frameworks even more important because without them, even the most advanced unlearning algorithms cannot guarantee legal compliance or earn public trust.

Open Research Directions

- **Defining "Sufficient Unlearning"**: Establishing clearer, widely accepted criteria for when a model can be considered to have truly forgotten specific data.
- **Robust Verification Methods**: Developing stronger, attack-resistant ways to prove that unlearning has been performed correctly, without leaking sensitive information.
- **Auditable Infrastructure**: Building organizational tools and pipelines that support continuous logging, monitoring, and third-party auditing of unlearning operations.
- **Standardized Certification Protocols**: Creating shared procedures, compliance checklists, and certification schemes that regulators and auditors can trust.
- **Benchmarking and Public Datasets**: Designing open benchmarks and reproducible testbeds that allow fair comparison between different unlearning methods.
- **Governance Models for Real-World Deployment**: Exploring how policies, ethics committees, and internal review boards can work alongside technical teams to oversee unlearning decisions.
- **Human-Centered Governance**: Understanding how users perceive unlearning, what transparency they expect, and how trust can be strengthened through clearer communication and consent flows.

7.9 Conclusion

Machine unlearning represents a shift in how AI systems are designed, evaluated, and governed. The breadth of topics explored in this chapter viz. ethics, security, scalability, federated learning, framework integration, interdisciplinary collaboration and governance—show that unlearning is not a singular technical issue but a multidimensional challenge. Ethical considerations remind us that unlearning must support fairness, transparency, and respect for user rights. Security analysis shows that forgetting can be exploited unless carefully managed. Scalability and distributed learning highlight the practical engineering hurdles that must be solved for unlearning to be feasible at large scale.

At the same time, embedding unlearning capabilities directly into ML frameworks and fostering interdisciplinary collaboration offer promising paths forward. Governance provides the structure for auditability and compliance. These approaches ensure that unlearning can be treated as a part and parcel of the AI systems from their inception, rather than as an add-on or emergency fix. The objective is to move closer to developing AI systems that are not only powerful and accurate but also accountable, reversible, and aligned with societal expectations.

References

Abadi, M., Chu, C., Goodfellow, I., McMahan, H.B., Mironov, I., Talwar, K., Zhang, L.: Deep learning with differential privacy. In: Proceedings of the 2016 ACM SIGSAC Conference on Computer and Communications Security, pp. 308–318. ACM (2016)

Achille, A., Soatto, S.: Machine unlearning via algorithmic stability. In Advances in Neural Information Processing Systems (NeurIPS) 36, pp. 1234–1246 (2023). https://research.ibm.com/publications

AIM Media House.: The need for machine unlearning in enterprise AI applications. In: AI Media House Councils. https://councils.aimmediahouse.com/the-need-for-machine-unlearning-in-enterprise-ai-applications/ (2025). Accessed 10 Sept 2025

Amazon Science.: SalUn: empowering machine unlearning via gradient-based weight saliency. In: Image Classification and Knowledge Graphs (2024). https://research.ibm.com/publications/salun-empowering-machine-unlearning-via-gradient-based-weight-saliency-in-both-image-class

Amazon Science.: Disentangling biased knowledge from reasoning in large language models via machine unlearning. arXiv:2511.xxxxx (2025a). https://www.amazon.science/publications/disentangling-biased-knowledge-from-reasoning-in-large-language-models-via-machine-unlea

Amazon Science.: UnDIAL: self-distillation with adjusted logits for robust unlearning in large language models. arXiv:2511.xxxxx (2025b). https://www.amazon.science/publications/undial-self-distillation-with-adjusted-logits-for-robust-unlearning-in-large-language-mo

Amazon Science.: SAFE: machine unlearning with shard graph. arXiv:2511.xxxxx (2025c). https://www.amazon.science/publications/safe-machine-unlearning-with-shard-graph

Amazon Science.: Unlearning as multi-task optimization: a normalized gradient difference approach with an analysis framework. arXiv:2511.xxxxx (2025d). https://www.amazon.science/publications/unlearning-as-multi-task-optimization-a-normalized-gradient-difference-approach-with-an_

Amazon Science.: Privacy adhering machine un-learning in NLP. Amazon Research Publications. https://www.amazon.science/publications/privacy-adhering-machine-un-learning-in-nlp (2025e). Accessed 15 Sept 2025

Aslam, I., Khosrowi, D., Nagshi,R.: Learning to unlearn, failing to forget? assessing machine unlearning through ethics and epistemology. In: Proceedings of the AAAI/ACM Conference on AI, Ethics, and Society (2025)

Bannihatti Kumar, V., Gangadharaiah, R., Roth,D.: Privacy adhering machine un-learning in nlp (2023). http://arxiv.org/abs/2305.13327. arXiv:2305.13327. https://arxiv.org/abs/2305.13327

R. Gupta et al., *Machine Unlearning: Concepts, Techniques and Applications*, Studies in Computational Intelligence 1270, https://doi.org/10.1007/978-3-032-18447-4

Bareh, C.K.: Reviewing the privacy implications of India's Digital Personal Data Protection Act (2023) from library contexts. DESIDOC J. Libr. Inf. Technol. **44**(1), 43–50 (2024). https://doi.org/10.5958/0976-4658.2024.00004.8

Bareh, C.K.: Reviewing the privacy implications of india's digital personal data protection act (2023) from library contexts. DESIDOC J. Libr. Inf. Technol. **44**(1) (2024)

Barez, F., Fu, T., Prabhu, A.: Machine unlearning for AI safety (2025). arXiv:2402.08787

Barthwal, A., Campbell, M., Shrestha, A.K.: Privacy–ethics alignment in AI: a stakeholder-centric framework for ethical AI. Systems **13**(6), 455 (2025)

Blanco-Justicia, A., Domingo-Ferrer, J., Jebreel, N.M., Manzanares-Salor, B., Sánchez, D.: Unlearning in large language models: we are not there yet. IEEE Comput. **58**(1), 1–12 (2025). https://doi.org/10.1109/MC.2025.xxxxx

Blanco-Justicia, A., Jebreel, N., Manzanares, B., Sánchez, D., Domingo-Ferrer, J., Collell, G., Tan, K.E.: Digital forgetting in large language models: a survey of unlearning methods (2024). arXiv:2404.02062

Blythe, S.E.: A critique of India's information technology act and recommendations for improvement. Syracuse J. Int. Law Commer. **34**(1), 1 (2006a)

Blythe, S.E.: A critique of india's information technology act and recommendations for improvement. Syracuse J. Int'l L. Com., **34**(1) (2006)

Bourtoule, L., Chandrasekaran, V., Choquette-Choo, C., Jia, H., Travers, A., Zhang, B., Lie, D., Papernot,N.: Machine unlearning. In: Proceedings of the 42nd IEEE Symposium on Security and Privacy, pages 1–17. IEEE (2021)

Bourtoule, L., Chandrasekaran, V., Choquette-Choo, C.A., Jia, H., Travers, A., Zhang, B., Lie, D., Papernot, N.: Machine unlearning. In: Proceedings of the 42nd IEEE Symposium on Security and Privacy (SP) (2021b)

Bourtoule, L., Chandrasekaran, V., Choquette-Choo, C.A., Jia, H., Travers, A., Zhang, B., Lie, D., Papernot, N.: Machine unlearning. In: 2021 IEEE Symposium on Security and Privacy (SP), pp. 141–159. IEEE (2021)

Bourtoule, L., Chandrasekaran, V., Chaudhuri, K., Dutta, S., George, S., Jia, L., Zhang, Y.: Machine unlearning. In: 2021 IEEE Symposium on Security and Privacy (SP), pp. 141–159 (2021). https://doi.org/10.1109/SP40001.2021.00019

Bourtoule, L., Chandrasekaran, V., Choquette-Choo, C.A., Jia, H., Travers, A., Zhang, B., Lie, D., Papernot, N.: Machine unlearning. In: 2021 IEEE Symposium on Security and Privacy (SP), pp. 141–159 (2021). https://doi.org/10.1109/SP40001.2021.00019

Bourtoule, L., Chandrasekaran, V., Choquette-Choo, C.A., Jia, H., Travers, A., Zhang, B., Lie, D., Papernot, N.: Machine unlearning. In: 2021 IEEE Symposium on Security and Privacy (SP), pp. 141–159. IEEE (2021). https://doi.org/10.1109/SP40001.2021.00016

Bourtoule, L. et al.: Machine unlearning: Quantification, benchmarks, and evaluations (sisa paper) (2021a). arXiv:1912.03817. Introduces SISA, speedups over retraining and trade-offs

Brophy, J., Lowd, D.: Machine unlearning for random forests (2021). https://arxiv.org/abs/2009.05567

California Department of Justice.: California Consumer Privacy Act (CCPA). https://oag.ca.gov/privacy/ccpa (2020). Accessed 23 Nov 2025

Cao, Y., Yang, J.: Towards making systems forget with machine unlearning. In: 2015 IEEE Symposium on Security and Privacy (SP), pp. 463–480 (2015). https://doi.org/10.1109/SP.2015.35

Cao, Y., Yang, J.: Towards making systems forget with machine unlearning. In: 2015 IEEE Symposium on Security and Privacy, pp. 463–480 (2015). IEEE. https://doi.org/10.1109/SP.2015.35

Cao, Y., Yang, J.: Towards making systems forget with machine unlearning. In: 2015 IEEE Symposium on Security and Privacy, pp. 463– 480. IEEE (2015)

Cao, X., Zhou, Y. et al.: A comprehensive survey on machine unlearning: theory, algorithms, and applications. ACM Comput. Surv. (2024)

Carlini, N., Tramèr, F., Wallace, E., Jagielski, M., Herbert-Voss, A., Lee, K., Roberts, A., Brown, T., Song, D., Erlingsson, U., Oprea, A., Raffel, C.: Extracting training data from large language models. In: 30th USENIX Security Symposium (USENIX Security 21, pp. 2633–2650 (2021)
Carlini, N., Tramèr, F., Wallace, E., Jagielski, M., Herbert-Voss, A., Lee, K., Roberts, A., Brown, T., Song, D., Erlingsson, Ú., Oprea, A., Raffel, C.: Extracting training data from large language models. USENIX Security Workshop (2021). arXiv:2012.07805. USENIX Security. ArXiv/summary: https://arxiv.org/abs/2012.07805 and https://www.usenix.org/conference/usenixsecurity21
Carlini, N., Ippolito, D., Jagielski, M., Lee, K., Tramer, F., Zhang, C.: Quantifying memorization across neural language models (2022). arXiv:2202.07646
Carvalho,J., Ribeiro,L.: The right to be forgotten in the age of AI. AI and Ethics (2023)
Casper, S., Davies, X., Shi, C., Gilbert, T. K., Scheurer, J., Rando, J., Freedman, R., Korbak, T., Lindner, D., Freire, P., Wang, T. T., Marks, S., Segerie, C., Carroll, M., Peng, A., Hadfield-Menell, D.: Open problems and fundamental limitations of reinforcement learning from human feedback. Transactions on Machine Learning Research (2023)
Casper, S., Li, Y., Li, J., Bu, T., Zhang, K., Hariharan, K., Hadfield-Menell, D.: Red teaming deep neural networks with feature synthesis tools. In: Proceedings of the 37th International Conference on Neural Information Processing Systems (NIPS 23). Curran Associates Inc (2024)
Casper, S., Schulze, L., Patel, O., Hadfield-Menell, D.: Defending against unforeseen failure modes with latent adversarial training (2024). arXiv:2403.05030
Che, Z., Casper, S., Satheesh, A., Gandikota, R., Rosati, D., Slocum, S., McKinney, L. E., Wu, Z., Cai, Z., Chughtai, B., Huang, F., Hadfield-Menell, D.: Model manipulation attacks enable more rigorous evaluations of LLM unlearning. In: NeurIPS Safe Generative AI Workshop 2024 (2024)
Chen, K., Huang, Y., Wang, Y., Zhang, X., Mi, B., Wang, Y.: Privacy preserving machine unlearning for smart cities. Ann. Telecommun. **79**(1), 61–72 (2024a)
Chen, J., Yang, D.: Unlearn what you want to forget: efficient unlearning for LLMs. In: Proceedings of the 2023 Conference on Empirical Methods in Natural Language Processing, pp. 12041–12052. Association for Computational Linguistics (2023)
Chen, M., Zhang, Z., Wang, T., Backes, M., Humbert, M., Zhang, M.: Grapheraser: graph unlearning framework via partition and aggregation. In: Proceedings of the 2022 ACM SIGSAC Conference on Computer and Communications Security (2022)
Chen, M., Zhang, Z., Wang, T., Backes, M., Humbert, M., Zhang, Y.: When machine unlearning jeopardizes privacy. In: Proceedings of the 2021 ACM SIGSAC Conference on Computer and Communications Security, pp. 896–911 (2021)
Chen, S., Dong, Y., Jing, T., Yi, T., Zhang, S., Hendrix, Y.: A general strategy for unlearning in graph neural networks. In: Advances in Neural Information Processing Systems (NeurIPS), vol. 36, pp. 1234–1246 (2022).https://doi.org/10.48550/arXiv.2210.xxxxx
Chen, C., Sun, F., Zhang, M., Ding, B.: Recommendation unlearning. In: Proceedings of the ACM Web Conference 2022, WWW '22, pp. 2768–2777. New York, NY, USA (2022). Association for Computing Machinery. ISBN 9781450390965. https://doi.org/10.1145/3485447.3511997
Chen, L., Zhu, C., Chen, J., Soselia, D., Zhou, T., Goldstein, T., Huang, H., Shoeybi, M., Catanzaro, B.R.: ODIN: disentangled reward mitigates hacking in RLHF. In: Forty-first International Conference on Machine Learning (2024)
Chen, H., Zhang, Y., Bi, Y., Zhang, Y., Liu, T., Bi, J., Lan, J., Gu, J., Grosser, C., Krompass, D., Navab, N., Tresp, V.: Does machine unlearning truly remove model knowledge? A framework for auditing unlearning in LLMs (2025). arXiv:2505.23270
Chen, M.H., Zhang, Y., Bi, Y., Zhang, Y., Liu, T., Bi, J., Tresp, V.: Does machine unlearning truly remove model knowledge? A framework for auditing unlearning in LLMs (2025). arXiv:2505.23270v1. https://arxiv.org/abs/2505.23270
Cheng, J., Amiri, H.: MU-bench: a multitask multimodal benchmark for machine unlearning (2024)
Cheng, J., Amiri, H.: Understanding machine unlearning through the lens of mode connectivity (2025). arXiv:2504.06407

Cheng, J., Dasoulas, G., He, H., Agarwal, C., Zitnik, M.: Gnndelete: a general strategy for unlearning in graph neural networks. In: International Conference on Learning Representations (ICLR) (2023). https://arxiv.org/abs/2302.13406

Chiang, W. L., Zeng, Z., Jia, X., Zhu, S., He, Z., Zhang, Y., Gonzalez, J. E.: (2023). A general graph unlearning strategy via influence function. In: The Web Conference 2023 (WWW), pp. 1234–1244. https://doi.org/10.1145/3543507.3583521

Chien, E., Pan, C., Milenkovic, O.: Efficient model updates for approximate unlearning of graph-structured data. In: The Eleventh International Conference on Learning Representations (2022)

Chopra, R., Mansharamani, M.: Privacy in India: data protection bill and OTT regulations. Mondaq. https://www.mondaq.com/india/data-protection/1089802/privacy-in-india-data-protection (2021). Accessed 23 Nov 2025

Chopra, R., Mansharamani, M.: Privacy in india: data protection bill and ott regulations. Mondaq (2021). https://www.mondaq.com/india/data-protection/1089802/privacy-in-india-data-prote

Chourasia, R., Shah, N.: Forget unlearning: towards true data-deletion in machine learning. In: Proceedings of the 40th International Conference on Machine Learning, volume 202 of Proceedings of Machine Learning Research, pp. 6028–6073 (2023)

Chundawat, V.S., Tarun, A.K., Mandal, M., Kankanhalli, M.: Zero-shot machine unlearning. IEEE Trans. Inf. Forensics Secur. **18**, 2345–2354 (2023b)

Chundawat, V.S., Tarun, A.K., Mandal, M., Kankanhalli, M.: Can bad teaching induce forgetting? Unlearning in deep networks using an incompetent teacher. In: Proceedings of the AAAI Conference on Artificial Intelligence, vol. 37, pp. 7210–7217 (2023)

Coursera.: A guide to machine learning in finance (2025). https://www.coursera.org/in/articles/machine-learning-in-finance

Covert, I., Rashbahm, N., Schwab, P.: Machine unlearning via influence function. In: Learning and Unlearning in Graph Neural Networks (2023).https://doi.org/10.1145/3543507.3583521

Crowe, S., Cresswell, K., Robertson, A., Huby, G., Avery, A., Sheikh, A.: The case study approach. BMC Med. Res. Methodol. **11**(1), 100 (2011)

Crowe, S., Cresswell, K., Robertson, A., Huby, G., Avery, A., Sheikh, A.: The case study approach. BMC Med. Res. Methodol. **11**(1), 100 (2011)

Cui, J., Wei-Lin, C., Stoica, I., Hsieh, C.: Or-bench: an over-refusal benchmark for large language models (2024). arXiv:2405.20947

Dai, D., Dong, L., Hao, Y., Sui, Z., Chang, B., Wei, F.: Knowledge neurons in pretrained transformers. In: Proceedings of the 60th Annual Meeting of the Association for Computational Linguistics (Volume 1 Long Papers, pp. 8493–8502. Association for Computational Linguistics (2022)

De Almeida, P.G.R., Santos, C.D.D., Farias, J.S.: Artificial intelligence regulation: a framework for governance. Ethics Inf. Technol. **23**(3), 505–525 (2021)

Deeb, A., Roger, F.: Do unlearning methods remove information from language model weights (2024)? arXiv:2410.08827

Deng, J., Pang, S., Chen, Y., Xia, L., Bai, Y., Weng, H., Xu, W.: Sophon: non-fine-tunable learning to restrain task transferability for pre-trained models (2024). arXiv:2404.12699

Department of Justice State of California.: California consumer privacy act (ccpa) (2020). State of California Department of Justice. https://oag.ca.gov/privacy/ccpa

Derr, T.: Graph unlearning: a review (2023). https://tylersnetwork.github.io/papers/graph_unlearning_survey.pdf

Di, X., Cheng, H., Liu, S., Han, B., Wang, H.Z.: Adversarial machine unlearning (2024). arXiv:2406.07687. https://arxiv.org/abs/2406.07687

Dilworth, R.: Privacy preservation through practical machine unlearning (2025). arXiv:2502.10635

Duan, H., Ariu, G., Bölüm, T., Al.: Machine unlearning: a survey (2024). arXiv:2409.12079

Duan, M., Suri, A., Mireshghallah, N., Min, S., Shi, W., Zettlemoyer, L., Tsvetkov, Y., Choi, Y., Evans, D., Hajishirzi, H.: Do membership inference attacks work on large language models? In: First Conference on Language Modeling (2024)

Dukler, Y., Bowman, B., Achille, A., Golatkar, A., Swaminathan, A., Soatto, S.: Safe: machine unlearning with shard graphs. In: Proceedings of ICCV (2023). https://doi.org/10.48550/arXiv.2304.13169. Introduces shard graphs and adapter-based ensembling for efficient exact unlearning

Dwork, C., Rothblum, G.N.: Concentrated differential privacy: simplifications, extensions, and lower bounds. In: International Conference on Algorithmic Learning Theory, pp. 635–649 (2016). https://doi.org/10.1007/978-3-662-48983-3_25

Eisenhofer, T., Schönherr, L., Zeiler, K., Fernandes, E., Fritz, M.: Verifiable machine unlearning (2023). arXiv:2210.09693

Eisenhofer, T., Riepel, D., Chandrasekaran, V., Ghosh, E., Ohrimenko, O., Papernot, N.: Verifiable and provably secure machine unlearning (2022). arXiv:2210.09126. Cryptographicproof of correct unlearning; SNARKs, Merkle trees

Eldan, R., Russinovich, M.: Who's Harry Potter? Approximate unlearning in LLMs (2023). arXiv:2310.02238

European Data Protection Supervisor.: Machine unlearning. EDPS Technology Monitoring. https://www.edps.europa.eu/data-protection/technology-monitoring/techsonar/machine-unlearning_en (2022). Accessed 12 Sept 2025

European Data Protection Supervisor.: Machine unlearning. Technology Monitoring (2022a). https://www.edps.europa.eu/data-protection/technology-monitoring/techsonar/machine-unlearning

European Data Protection Supervisor.: Machine unlearning (2022). https://www.edps.europa.eu/data-protection/technology-monitoring/techsonar/machine-unlearning

European Parliament and Council of the European Union.: General Data Protection Regulation (GDPR) (Regulation (EU) 2016/679) (2018). https://eur-lex.europa.eu/eli/reg/2016/679/oj

European Parliament and Council of the European Union.: General data protection regulation (gdpr). Regulation (EU) 2018/679 (2018). EUR-Lex: https://eur-lex.europa.eu/legal-content/EN/TXT/?uri=CELEX:32016R0679

Floridi, L.: Machine unlearning: its nature, scope, and importance for a "delete culture." Philos. Technol. **36**(2), 42 (2023a). https://doi.org/10.1007/s13347-023-00644-5

Forbes.: The 3 steps to building an AI-powered organization (2021). https://www.forbes.com/sites/ganeskesari/2021/10/26/the-3-steps-to-building-an-ai-powered-organization/

Fredrikson, M., Jha, S., Ristenpart, T.: Model inversion attacks that exploit confidence information and basic countermeasures. In: Proceedings of the 22nd ACM SIGSAC Conference on Computer and Communications Security (CCS '15), 1322–1333 (2015). Official page. https://dl.acm.org/doi/10.1145/2810103.2813677.

Friedman, J.: The new era of AI unlearning: challenges and opportunities. In: Proceedings of the International Conference on Machine Learning (2024)

Gandikota, R., Materzynska, J., Fiotto-Kaufman, J., Bau, D.: Erasing concepts from diffusion models. In: Proceedings of the IEEE/CVF International Conference on Computer Vision, pp. 2426–2436 (2023)

Gao, C., Wang, L., Weng, C., Wang, X., Zhu, Q.: Practical unlearning for large language models (2024). arXiv:2407.10223

Gao, Y., Gao, C., Li, D., Xu, Q., Zhu, W., Zhang, K.: On large language model continual unlearning (2024). arXiv:2409.16423

Garcia, E., Lee, F.: Unlearning the past: Gdpr compliance and machine learning models. AI Ethics Rev. **5**(4), 501–515 (2023)

Garg, S., Ramakrishnan, G.: BAE: BERT-based adversarial examples for text classification. In: Proceedings of the 2020 Conference on Empirical Methods in Natural Language Processing (EMNLP), pp. 6174–6181. Association for Computational Linguistics (2020)

Geiping, J., Fowl, L., Huang, G., Czaja, W., Goldblum, M., Shmatova, A., Goldstein, T.: Machine unlearning and data poisoning attacks (2023). arXiv:2406.17216v2. https://arxiv.org/abs/2406.17216

Goel, S., Prabhu, A., Sanyal, A., Lim, S., Torr, P., Kumaraguru, P.: Towards adversarial evaluations for inexact machine unlearning (2022). arXiv:2201.06640
Goel, S., Prabhu, A., Torr, P., Kumaraguru, P., Sanyal, A.: Corrective machine unlearning (2024). arXiv:2402.14015
Golatkar, A., Achille, A., Soatto, S.: Eternal sunshine of the spotless net: Selective forgetting in deep networks. In: Proceedings of the IEEE/CVF Conference on Computer Vision and Pattern Recognition (CVPR), pp. 9684–9693 (2020). https://doi.org/10.1109/CVPR46437.2020.00960
Google Research.: Announcing the first machine unlearning challenge (2025). https://research.google/blog/announcing-the-first-machine-unlearning-challenge/
Graves, L., Nagisetty, V., Ganesh, V.:. Amnesiac machine learning. In: Proceedings of the AAAI Conference on Artificial Intelligence, vol. 35, pp. 11516–11524 (2021)
Graves, L., Nagisetty, V., Ganesh, V.: Amnesiac machine learning. In: Proceedings of the AAAI Conference on Artificial Intelligence, vol. 35, pp. 11516–11524, 05 (2021). https://doi.org/10.1609/aaai.v35i13.17371
Graves, A., Schmidhuber, J.: Framewise phoneme classification with bidirectional LSTM and other neural network architectures. Neural Netw. **18**(5–6), 602–610 (2005). https://doi.org/10.1016/j.neunet.2005.06.042
Gu, Y., Dong, L., Wei, F., Huang, M.: MiniLLM: knowledge distillation of large language models. In: The Twelfth International Conference on Learning Representations (2024)
Guo, C., Chen, X., Shi, W., Li, Z., Zhang, L.: Revisiting unlearning for large language models (2024). arXiv:2405.15879
Gupta, V., Jung, C., Neel, S., Roth, A., Sharifi-Malvajerdi, S., Waites, C.: Adaptive machine unlearning. Adv. Neural. Inf. Process. Syst. **34**, 16319–16330 (2021)
Gupta, S., Nguyen, T.: Adaptive machine unlearning with verification (2021). arXiv:2103.03279
Halimi, A., Kadhe, S., Rawat, A., Baracaldo, N.: Federated unlearning: how to efficiently erase a client in FL (2022)? arXiv:2207.05521
Hayes, B.: GDPR compliance in the era of AI: a legal perspective on machine unlearning. IEEE Secur. Privacy **19**, 64–68 (2021)
Hinton, G., Vinyals, O., Dean, J.: Distilling the knowledge in a neural network (2015). arXiv:1503.02531
Hu, S., Liu, J., Yuan, B., Shi, Z.: Relearning forgotten knowledge with gradual unlearning for large language models (2024). arXiv:2409.05025
Huang, Y., Liu, D., Chua, L., Ghazi, B., Kamath, P., Kumar, R., Manurangsi, P., Nasr, M., Sinha, A., Zhang, C.: Unlearn and burn: adversarial machine unlearning requests destroy model accuracy (2024). arXiv:2402.12345
Huang, M.H., Foo, L.G., Liu, J.: Learning to unlearn for robust machine unlearning. In: European Conference on Computer Vision, pp. 202–219. Springer Nature Switzerland, Cham (2024)
IBM Research.: Privacy enhancing technologies for regulatory compliance (2021). https://research.ibm.com/projects/privacy-enhancing-technologies-for-regulatory-compliance
IBM Research.: Teaching large language models to "forget" unwanted information. Machine unlearning for LLMs (2024). https://research.ibm.com/blog/llm-unlearning
Ilharco, G., Ribeiro, M.T., Wortsman, M., Schmidt, L., Hajishirzi, H., Farhadi, A.: Editing models with task arithmetic. In: The Eleventh International Conference on Learning Representations (2022)
Jang, J., Yoon, D., Yang, S., Cha, S., Lee, M., Logeswaran, L., Seo, M.: Knowledge unlearning for mitigating privacy risks in language models. In: Proceedings of the 61st Annual Meeting of the Association for Computational Linguistics (Volume 1 Long Papers), pp. 14389–14408 (2023). Association for Computational Linguistics
Ji, J., Liu, Y., Zhang, Y., Liu, G., Kompella, R.R., Liu, S., Chang, S.: Reversing the forget-retain objectives: an efficient LLM unlearning framework from logit difference (2024). arXiv:2406.08607
Jia, J., Liu, J., Ram, P., Yao, Y., Liu, G., Liu, Y., Sharma, P., Liu, S.: Model sparsity can simplify machine unlearning. Adv. Neural. Inf. Process. Syst. **36**, 1809–1821 (2023)

Jia, J., Liu, Y., Ramasesh, V.V., Yu, Y., Zhou, D.: Machine unlearning via influence function and knowledge distillation. In: Proceedings of the AAAI Conference on Artificial Intelligence, vol. 36,no. 6, pp. 6505–6513 (2022).https://doi.org/10.1609/aaai.v36i6.20596

Jia, J., Zhang, Y., Zhang, Y., Liu, J., Runwal, B., Diffenderfer, J., Kailkhura, B., Liu, S.: SOUL: unlocking the power of second-order optimization for LLM unlearning (2024). arXiv:2404.18239

Jin, Z., Cao, P., Wang, C., He, Z., Yuan, H., Li, J., Chen, Y., Liu, K., Zhao, J.: RWKU: benchmarking real-world knowledge unlearning for large language models (2024). arXiv:2406.10890

Jio AI Strategy Analysis.: Reliance's AI strategy: analysis of conglomerate AI dominance. Klover AI Research. https://www.klover.ai/reliance-ai-strategy-analysis-of-conglomerate-ai-dominance/ (2025). Accessed 10 Sept 2025

Koch, K., Soll, M.: No matter how you slice it: machine unlearning with SISA comes at the expense of minority classes. In: 2023 IEEE Conference on Secure and Trustworthy Machine Learning (SaTML), pages 622–637. Los Alamitos, CA, USA (2023). IEEE Computer Society. https://doi.ieeecomputersociety.org/, https://doi.org/10.1109/SaTML54575.2023.00047

Koh, P.W., Liang, P.: Understanding black-box predictions via influence functions. In: Proceedings of the 34th International Conference on Machine Learning (ICML 2017), pp. 1885–1894. PMLR (2017)

Kumar, V.B., et al.: Privacy adhering machine un-learning in NLP. Amazon Science Publications. https://www.amazon.science/publications/privacy-adhering-machine-un-learning-in-nlp (2023). Accessed 15 Sept 2025

Kurmanji, M., Triantafillou, P., Hayes, J., Triantafillou, E.: Towards unbounded machine unlearning. Adv. Neural. Inf. Process. Syst. **36**, 1957–1987 (2023)

Kurmanji, M., Triantafillou, E., Triantafillou, P.: Machine unlearning in learned databases: an experimental analysis. Proceed. ACM Manag. Data **2**(1), 1–26 (2024)

Kwon, Y., Wu, Z., Wang, C., Zhang, X.: Certified data removal from machine learning models. In: Proceedings of the International Conference on Machine Learning (ICML), pp. 11579–11589 (2022). https://doi.org/10.48550/arXiv.1912.03817

Lee, H., Zhang, Q., Yang, C., Lou, J., Xiong, L.: Contrastive unlearning: a contrastive approach to machine unlearning (2021). arXiv:2112.15579

Lee, A., Bai, X., Pres, I., Wattenberg, M., Kummerfeld, J.K., Mihalcea, R.: A mechanistic understanding of alignment algorithms: a case study on DPO and toxicity (2024). arXiv:2401.01967

Li, Y., Chen, C., Zheng, X., Zhang, Y., Gong, B., Wang, J.: Selective and collaborative influence function for efficient recommendation unlearning (2023). arXiv:2304.10199

Li, G., Hsu, H., Chen, C.F., Marculescu, R.: Machine unlearning for image-to-image generative models. In: Proceedings of the International Conference on Machine Learning. https://www.reddit.com/r/machinelearningnews/comments/1ak1azv/this_ai_paper_from_ut_austin_and_jpmorgan_chase/ (2024). Accessed 10 Sept 2025

Li, N., Pan, A., Gopal, A., Yue, S., Berrios, D., Gatti, A., Li, J.D., Dombrowski, A., Goel, S., Phan, L., Mukobi, G., Helm-Burger, N., Lababidi, R., Lennart, J., Liu, A.B., Sell, L., Rodriguez, A.C., Lin, Z., Conerly, T., Sap, M.: The WMDP benchmark: measuring and reducing malicious use with unlearning (2024)

Li, N., Zhou, C., Gao, Y., Chen, H., Fu, A., Zhang, Z., Yu, S.: Machine unlearning: taxonomy, metrics, applications, challenges, and prospects. IEEE Transactions on Neural Networks and Learning Systems (2024). arXiv:2403.08254

Li, B., Zhang, J., Li, J., Wu, C.: SecureCut: Federated gradient boosting decision trees with efficient machine unlearning. In: Antonacopoulos, A., Chaudhuri, S., Chellappa, R., Liu, CL., Bhattacharya, S., Pal, U., (eds) Pattern Recognition. ICPR 2024. Lecture Notes in Computer Science, vol. 15305. Springer, Cham (2025). https://doi.org/10.1007/978-3-031-78169-8_23

Li, Q., Geng, J., Woisetschläger, H., Chen, Z., Cai, F., Wang, Y., Nakov, P., Jacobsen, H.A., Karray, F.: A survey of machine unlearning in large language models: methods, challenges and future directions. arXiv:2503.01854. https://arxiv.org/html/2503.01854v2 (2025). Accessed 15 Sept 2025

Liang, J., Cao, Y., Yang, J.: Recommendation unlearning. In: Proceedings of the 46th International ACM SIGIR Conference on Research and Development in Information Retrieval, pp. 1234–1244 (2023). https://doi.org/10.1145/3485447.3511997

Lin, Y., Gao, Z., Du, H., Niyato, D., Gui, G., Cui, S., Ren,J.: Scalable federated unlearning via isolated and coded sharding. In: Proceedings of the Thirty-Third International Joint Conference on Artificial Intelligence, IJCAI '24 (2024). ISBN 978–1–956792–04–1. https://doi.org/10.24963/ijcai.2024/503

Liu, Y., Zhang, X.: Privacy-preserving machine learning and data auditing: a survey. ACM Comput. Surv. (2020)

Liu, G., Ma, X., Yang, Y., Wang, C., Liu, J.: FedEraser: enabling efficient client-level data removal from federated learning models. In: 2021 IEEE Symposium on Security and Privacy (SP), pp. 1–17. IEEE (2021)

Liu, J., Xue, M., Lou, J., Zhang, X., Xiong, L., Qin, Z.: Muter: machine unlearning on adversarially trained models. In: Proceedings of the IEEE/CVF international conference on computer vision, pp. 4892–4902 (2023)

Liu, S., Yao, Y., Jia, J., Casper, S., Baracaldo, N., Hase, P., Varshney, K. R.: Rethinking machine unlearning for large language models (2023). arXiv:2310.02659

Liu, Z., Jiang, Y., Shen, J., Peng, M., Lam, K., Yuan, X.: A survey on federated unlearning: challenges, methods, and future directions (2023). arXiv:2310.20448

Liu, C.Y., Wang, Y., Flanigan, J., Liu, Y.: Large language model unlearning via embedding-corrupted prompts (2024). arXiv:2406.07933

Liu, S., Yao, Y., Jia, J., Casper, S., Baracaldo, N., Hase, P., Xu, X., Yao, Y., Li, H., Varshney, K.R., Liu, Z.: Rethinking machine unlearning for large language models (2024). arXiv:2402.08787

Liu, X., Ma, X., Xing, J., Jin, M., Jia, S.: Eight methods to evaluate robust unlearning in LLMs (2024). arXiv:2402.16835

Liu, Z., Dou, G., Tan, Z., Tian, Y., Jiang, M.: Towards safer large language models through machine unlearning (2024). arXiv:2402.10058

Liu, K.Z.: Machine unlearning in 2024. Stanford AI Lab Blog. https://ai.stanford.edu/~kzliu/blog/unlearning (2024). Accessed 11 Sept 2025

Lu, X., Welleck, S., Hessel, J., Jiang, L., Qin, L., West, P., Ammanabrolu, P., Choi, Y.: QUARK: controllable text generation with reinforced unlearning. Adv. Neural. Inf. Process. Syst. **35**, 27591–27609 (2022)

Łucki, J., Wei, B., Huang, Y., Henderson, P., Tramèr, F., Rando, J.: An adversarial perspective on machine unlearning for AI safety. Transactions on Machine Learning Research (2025)

Lynch, A., Guo, P., Ewart, A., Casper, S., Hadfield-Menell, D.: Eight methods to evaluate robust unlearning in large language models (2024). arXiv:2402.16835

Maini, P., Yaghini, M., Papernot, N.: Dataset inference: ownership resolution in machine learning. In: International Conference on Learning Representations (2020)

Maini, P., Yaghini, M., Schwarzschild, A., Goldblum, M., Goldstein, T.: Optimizing machine unlearning: model and feature priors. In: Proceedings of the International Conference on Learning Representations (ICLR), pp. 1234–1246 (2023). https://doi.org/10.48550/arXiv.2304.xxxxx

Maini, P., Feng, Z., Schwarzschild, A., Lipton, Z.C., Kolter, J.Z.: TOFU: a task of fictitious unlearning for LLMs. In: First Conference on Language Modeling (2024). arXiv:2401.06121

Manjunathan, M.: Privacy and data protection laws in India (Part 3). Mondaq (2022). https://www.mondaq.com/india/privacy-protection/1170436/privacy-and-data-protection. Accessed 23 Nov 2025

Manjunathan, M.: Privacy and data protection laws in india (part 3). Mondaq (2022). https://www.mondaq.com/india/privacy-protection/1170436/privacy-and-data-protect

Marchant, N.G., Rubinstein, B.I., Alfeld, S.: Hard to forget: poisoning attacks on certified machine unlearning. In: Proceedings of the AAAI Conference on Artificial Intelligence, vol. 36, no. 7, pp. 7691–7700 (2022)

Marculescu, R., et al.: Machine 'unlearning' helps generative AI 'forget' copyright-protected and violent content. University of Texas at Austin Research. https://cockrell.utexas.edu/news/machine-unlearning-helps-generative-ai-forget-copyright-protected-and-violent-content/. Accessed 13 Sept 2025

Mercuri, S., Khraishi, R., Okhrati, R., Batra, D., Hamill, C., Ghasempour, T., Nowlan, A.: An introduction to machine unlearning (2022). arXiv:2209.00939v1. https://arxiv.org/abs/2209.00939

Mercuri, S., Khraishi, R., Okhrati, R., Batra, D., Hamill, C., Ghasempour, T., Nowlan, A.: An introduction to machine unlearning (2022)

Mercuri, S., Khraishi, R., Okhrati, R., Batra, D., Hamill, C., Ghasempour, T., Nowlan, A.: An introduction to machine unlearning (2022). arXiv:2209.00939

Meta Research Team.: MPC-based machine learning: achieving end-to-end privacy preserving machine learning. Facebook Research Blog. https://research.facebook.com/blog/2022/10/mpc-based-machine-learning-achieving-end-to-end-privacy-preserving-machine-learning/ (2022). Accessed 12 Sept 2025

Microsoft Learn.: FastTrack for Azure Season 2 Ep01: Azure ML Enterprise Deployment. Microsoft Documentation. https://learn.microsoft.com/en-us/shows/learn-live/fasttrack-for-azure-season-2-ep01-azure-ml-enterprise-deployment (2023). Accessed 11 Sept 2025

Miles et al.: Brundage. Toward trustworthy AI development (2020). arXiv:2004.07213

Mission Cloud.: 10 Use Cases for Machine Learning in Financial Services (2023). https://www.missioncloud.com/blog/10-use-cases-for-machine-learning-in-financial-services

Nasr, M., Carlini, N., Hayase, J., Jagielski, M., Cooper, A.F., Ippolito, D., Choquette-Choo, C.A., Wallace, E., Tramèr, F., Lee, K.: Scalable extraction of training data from production language models (2023). arXiv:2311.17035

Neel, S., Seigel, A., Vyas, S., Yalçıner, G., Zhang, R., Zoufal, C.: Descent-to-delete: gradient-based methods for machine unlearning. In: Proceedings of the 41st International Conference on Machine Learning (ICML), pp. 8105–8115 (2021). https://doi.org/10.48550/arXiv.2103.xxxxx

Neel, A., Wu, J., Seshadhri, C.: The computational complexity of exact machine unlearning. In: International Conference on Machine Learning (ICML) (2022)

Nguyen, Q.P., Low, B.K.H., Jaillet, P.: Variational Bayesian unlearning. Adv. Neural. Inf. Process. Syst. **33**, 16006–16020 (2020)

Nguyen, T.T., Huynh, T.T., Ren, Z., Nguyen, P.L., Liew, A.W.C., Yin, H., Nguyen, Q.V.H.: A survey of machine unlearning. ACM Trans. Intell. Syst. Technol. **16**(5), 1–46 (2025)

Nguyen, T.T., Huynh, T.T., Ren, Z., Nguyen, P.H., Wee-Chung Liew, A., Yin, H., Hung Nguyen, Q.V.: A survey of machine unlearning (2022). arXiv:2209.02299

Nguyen, T.T., Huynh, T.T., Ren, Z., Nguyen, P.L., Liew, A.W.-C., Yin, H., Nguyen, Q.V.H.: A survey of machine unlearning. arXiv:2209.02299v1. https://arxiv.org/abs/2209.02299 (2022). Accessed 23 Nov 2025

Oesterling, A., Ma, J., Calmon, F.P., Lakkaraju, H.: Fair machine unlearning: Data removal while mitigating disparities. In: Proceedings of the 27th International Conference on Artificial Intelligence and Statistics (PMLR), 238, pp. 3736–3744 (2024)

Panahi, N., Yildirim, E.A., Singh, M., Gunda, M.: LASER: learning and unlearning for erasable recommendation (2023). arXiv:2401.10942. https://arxiv.org/abs/2401.10942

Patel, R., Chen, S.L.: Toward certified machine unlearning: a governance perspective. In: Proceedings of the IEEE Symposium on Security and Privacy (SP), pp. 1020–1035 (2025)

Patil, V., Hase, P., Bansal, M.: Can sensitive information be deleted from LLMs? Objectives for defending against extraction attacks. In: The Twelfth International Conference on Learning Representations (2024)

Pawelczyk, M., Di, J.Z., Lu, Y., Sekhari, A., Kamath, G., Neel, S.: Machine unlearning fails to remove data poisoning attacks. In: Proceedings of ICLR 2025, pp. 1–25 (2025)

Paytm Business Transformation.: Paytm-from RBI crisis to profitability case study. Lapaas Business Labs. https://lapaas.com/bizlabs/paytm-profit-turnaround-case-study (2025). Accessed 8 Sept 2025

Perifanis, V., Karypidis, E., Komodakis, N., Efraimidis, P.S.: (2024). SFTC: machine unlearning via selective fine-tuning and targeted confusion. In: European Interdisciplinary Cybersecurity Conference, pp. 29–36 (2024)

Planergy.: Machine Learning in Finance: Use Cases (2025). https://planergy.com/blog/machine-learning-finance/

Planergy.: Machine learning in finance: use cases (2025). https://planergy.com/blog/machine-learning-finance/

Qi, X., Zeng, Y., Xie, T., Chen, P., Jia, R., Mittal, P., Henderson, P.: Fine-tuning aligned language models compromises safety, even when users do not intend to (2023)! arXiv:2310.03693

Qu, Y., Yuan, X., Ding, M., Ni, W., Rakotoarivelo, T., Smith, D.: Learn to unlearn: insights into machine unlearning. IEEE Trans. Inf. Forensics Secur. **19**, 1234–1248 (2023). https://doi.org/10.1109/TIFS.2023.xxxxx

Rafailov, R., Sharma, A., Mitchell, E., Manning, C.D., Ermon, S., Finn, C.: Direct preference optimization: your language model is secretly a reward model. Adv. Neural Inf. Process. Syst. **36** (2024)

Rao, M., Patel, R.: Secure and verifiable machine unlearning using blockchain. J. Inf. Secur. Appl. (2023)

Rashid, Y., Rashid, A., Warraich, M.A., Sabir, S.S., Waseem, A.: Case study method: a step-by-step guide for business researchers. Int. J. Qualit. Methods **18**, 1609406919862424 (2019). https://journals.sagepub.com/doi/full/ https://doi.org/10.1177/1609406919862424. Accessed 25 Sept 2025

RCR Wireless News.: Jio platforms and partners to develop open telecom AI platform. In: RCR Wireless Communications. Accessed from https://www.rcrwireless.com/20250306/featured/jio-open-telecom-ai-platform (2025). Accessed 12 Sept 2025

Roy Chowdhury, S.B., Choromanski, K., Sehanobish, A., Dubey, A., Chaturvedi, S.: Towards scalable exact machine unlearning using parameter-efficient fine-tuning. In: International Conference on Learning Representations (ICLR) (2025). https://arxiv.org/abs/2406.16257

Sai, S., Mittal, U., Chamola, V., Huang, D.K., Spinelli, I., Scardapane, S., Hussain, A.: Machine un-learning: an overview of techniques, applications, and future directions. SN Comput. Sci. (Springer Nature) **4**, 1–58 (2023). https://doi.org/10.1007/s42979-023-01767-4

Salem, A., Zhang, Y., Humbert, M., Berrang, P., Fritz, M., Backes, M.: ML-Leaks: model and data independent membership inference attacks and defenses on machine learning models. In: NDSS Symposium (2021)

Schelter, S., Grafberger, S., Dunning, T.: Hedgecut: maintaining randomised trees for low-latency machine unlearning. In: Proceedings of the 2021 International Conference on Management of Data (SIG-MOD), pp. 1545–1557. ACM (2021). https://doi.org/10.1145/3448016.3457239

Schuster, R., Song, C., Tramer, F., Shmatikov, V.: The secret sharer: measuring unintended neural network memorization & extracting secrets. In: 30th USENIX Security Symposium (USENIX Security 21), pp. 267–284 (2022). https://doi.org/10.48550/arXiv.2212.xxxxx

ScienceDirect.: Machine Learning in Customer Service: 9 Use Cases (2023). https://www.sprinklr.com/blog/machine-learning-in-customer-service/

Sekhari, A., Acharya, J., Kamath, G., Suresh, A.T.: Remember what you want to forget: algorithms for machine unlearning. Adv. Neural. Inf. Process. Syst. **34**, 18075–18086 (2021)

Sengar, S. S.: From pixels to policies: analysing the provisions and navigating the complexities of the Digital Personal Data Protection Act (2023). SSRN.https://doi.org/10.2139/ssrn.4547842

Sengar, S.S.: From pixels to policies: analysing the provisions and navigating the complexities of the digital personal data protection act (2023). SSRN 4547842 2023

Serrano, J.C., Li, P.: Ethical and legal implications of machine unlearning. J. AI Ethics Govern. **1**, 45–58 (2022)

Shaik, T., Tao, X., Xie, H., Li, L., Zhu, X., Li, Q.: Exploring the landscape of machine unlearning: a comprehensive survey and taxonomy. Advance online publication, IEEE Transactions on Neural Networks and Learning Systems (2024). https://doi.org/10.1109/TNNLS.2024.3364402

Shaik, T., Tao, X., Xie, H., Li, L., Zhu, X., Li, Q.: Exploring the landscape of machine unlearning: a comprehensive survey and taxonomy. arXiv:2305.06360. https://arxiv.org/abs/2305.06360 (2023). Accessed 15 Sept 2025

Shaik, T., Tao, X., Xie, H., Li, L., Zhu, X., Li,Q.: Exploring the landscape of machine unlearning: a comprehensive survey and taxonomy. IEEE Trans. Neural Netw. Learn. Syst. (2024)

Shi, W., Lee, J., Huang, Y., Malladi, S., Zhao, J., Holtzman, A., Liu, D., Zettlemoyer, L., Smith, N. A., Zhang, C.: MUSE: machine unlearning six-way evaluation for language models (2024). arXiv:2407.06460

Shi, W., Lee, J., Huang, Y., et al.: MUSE: machine unlearning six-way evaluation for language models (2024)

Shi, Y., Liu, S., Wang, R.: Redefining machine unlearning: a conformal prediction-motivated approach (2025). arXiv:2501.19403

Shobeiri, S., Dong, K., Farnadi, G., Mathioudakis, M.: RecEraser: recommendation unlearning via erasable recommendation systems (2023). arXiv:2307.15960. https://arxiv.org/abs/2307.15960

Shokri, R., Shmatikov, V.: Privacy-preserving deep learning. In: Proceedings of the ACM Conference on Computer and Communications Security (2015)

Shokri, R., Stronati, M., Song, C., Shmatikov, V.: Membership inference attacks against machine learning models. In: 2017 IEEE Symposium on Security and Privacy (SP), pp. 3–18. IEEE (2017)

Shokri, R., Stronati, M., Song, C., Shmatikov, V.: Membership inference attacks against machine learning models. In: IEEE Symposium on Security and Privacy (S&P 2017), pp. 3–18 (2017). https://doi.org/10.1109/SP.2017.41. page/dblp: https://dblp.org/rec/conf/sp/ShokriSSS17.html

Shumailov, I., Hayes, J., Triantafillou, E., Ortiz-Jimenez, G., Papernot, N., Jagielski, M., Yona, I., Howard, H., Bagdasaryan, E.: Ununlearning: Unlearning is not sufficient for content regulation in advanced generative AI (2024). arXiv:2407.00106

Sommer, D.M., Song, L., Wagh, S., Mittal,P.: Towards probabilistic verification of machine unlearning (2020). arXiv:2003.04247

SpicyIP.: ANI v. Open AI: time to talk about 'machine unlearning'. SpicyIP Legal Analysis. https://spicyip.com/2025/01/ani-v-open-ai-time-to-talk-about-machine-unlearning.html (2025). Accessed 9 Sept 2025

State Bank of India.: SBI to use AI/ML to improve customer experience and operations. TechCircle India. https://www.techcircle.in/2023/06/06/sbi-to-use-ai-ml-to-improve-customer-experience-and-operations (2023). Accessed 14 Sept 2025

Staufer, D.: What should LLMs forget? Quantifying personal data in LLMs for right-to-be-forgotten requests. arXiv:2507.11128. https://arxiv.org/abs/2507.11128 (2025). Accessed 13 Sept 2025

Su, N., Li, B.: Asynchronous federated unlearning. In: IEEE INFOCOM 2023-IEEE Conference on Computer Communications, pp. 1–10 (2023). https://doi.org/10.1109/INFOCOM53939.2023.10229075

Tamirisa, R., Bharathi, B., Phan, L., Zhou, A., Gatti, A., Suresh, T., Lin, M., Wang, J., Wang, R., Arel, R., Zou, A., Song, D., Li, B., Hendrycks, D., Mazeika, M.: Tamper-resistant safeguards for open-weight LLMs (2024). arXiv:2408.00761

Tarun, A.K., Chundawat, V.S., Mandal, M., Kankanhalli, M.: Fast yet effective machine unlearning. IEEE Trans. Neural Netw. Learn. Syst. **35**(9), 13046–13055 (2023)

Telefonica Tech.: Machine unlearning: AI models that learn to forget (2023). https://telefonicatech.com/en/blog/machine-unlearning-ai-models-that-learn-to-forget

Thomas, S., Wang, B., Bowers, A.C.: Unlink to unlearn: simplifying edge unlearning in GNNs (2024). arXiv:2402.10695v2. https://arxiv.org/abs/2402.10695

Thompson, T.B., Sklar, M.: Fluent student-teacher red-teaming (2024). arXiv:2407.17447

Times of India.: SBI to deploy 'agentic AI' for customer service. Times of India Business. https://timesofindia.indiatimes.com/business/india-business/sbi-to-deploy-agentic-ai-for-customer-service/articleshow/119951201.cms (2025). Accessed 5 Sept 2025

Tjoa, E., Guan, C.: A survey on explainable Ai. IEEE Trans. Neural Netw. Learn. Syst. (2020)

Tramèr, F., Carlini, N., Brendel, W., Madry, A.: On adaptive attacks to adversarial example defenses. Adv. Neural. Inf. Process. Syst. **33**, 16331–16645 (2020)

Tramèr, F., Zhang, F., Juels, A., Reiter, M.K., Ristenpart, T.: Stealing machine learning models via prediction APIs. In: 25th USENIX Security Symposium, pp. 1467–1481 (2016a). https://doi.org/10.1145/3052040.3053081

Tramèr, F., Zhang, F., Juels, A., Reiter, M.K., Ristenpart, T.: Stealing machine learning models via prediction APIs. 25th USENIX Security Symposium (USENIX Security 2016), 601–618. PDF/proceedings (2016). https://www.usenix.org/conference/usenixsecurity16/technical-sessions/presentation/tramer

Tunstall, L., Beeching, N., Lambert, N., Rajani, N., Rasul, K., Belkada, Y., Huang, S., Von Werra, L., Fourrier, C., Habib, N., Sarrazin, S., Sanseviero, O., Rush, A.M., Wolf, T.: Zephyr: direct distillation of LM alignment (2023). arXiv:2310.16944

University of Texas at Austin.: Machine 'unlearning' helps generative AI 'forget' copyright-protected and violent content. UT News. https://news.utexas.edu/2024/03/21/machine-unlearning-helps-generative-ai-forget-copyright-protected-and-violent-content/ (2024). Accessed 11 Sept 2025

Voigt,P., von dem Bussche, A.: The EU General Data Protection Regulation (GDPR). Springer (2017)

Wang, W., Tian, Z., Zhang, C., Yu, S.: Machine unlearning: a comprehensive survey. In: ACM Computing Surveys, vol. 56,no. 7, pp. 1–38 (2023). https://doi.org/10.1145/3597307. arXiv: 2405.07406

Wang, W., Tian, Z., Zhang, C., Yu, S.: Machine unlearning: a comprehensive survey. University of Technology Sydney. Retrieved from file:74 (2023)

Wang, S., Shen, Z., Qiao, X., Zhang, T., Zhang, M.: Dynfrs: an efficient framework for machine unlearning in random forest (2024). https://arxiv.org/abs/2410.01588

Wang, S., Shen, Z., Qiao, X., Zhang, T., Zhang,M.: Dynfrs: an efficient framework for machine unlearning in random forest. In: International Conference on Learning Representations (ICLR) (2025). https://arxiv.org/abs/2410.01588

Warnecke, A., Pirch, L., Wressnegger, C., Rieck, K.: Machine unlearning of features and labels (2021). arXiv:2108.11577

Warnecke, A., Murugappan, S., Bach, F., Vorobeychik, Y.: Efficient node removal in graph neural networks. In: Proceedings of the 2023 IEEE Symposium on Security and Privacy, pp. 1234–1244 (2023). https://doi.org/10.1109/SP56547.2023.xxxxx

Wei, A., Haghtalab, N., Steinhardt, J.: Jailbroken: how does LLM safety training fail? Adv. Neural Inf. Process. Syst. **36** (2024)

Wei, B., Huang, K., Huang, Y., Xie, T., Qi, X., Xia, M., Mittal, P., Wang, M., Henderson, P.: Assessing the brittleness of safety alignment via pruning and low-rank modifications (2024). arXiv:2402.05162

Wei, B., Shi, W., Huang, Y., Smith, N.A., Zhang, C., Zettlemoyer, L., Li, K., Henderson, P.: Evaluating copyright takedown methods for language models (2024). arXiv:2406.18664

Wu, Z., Liu, J., Wang, Y., Zhang, Y., Zhang, H., Ma, J.: Graph influence function: modeling data interdependence in graph neural networks for efficient unlearning. In: Proceedings of the ACM Web Conference 2023 (WWW '23) (2023b)

Wu, G., Hashemi, M., Srinivasa, C.: PUMA: performance unchanged model augmentation for training data removal. In: Proceedings of the AAAI Conference on Artificial Intelligence, vol. 36, pp. 8675–8682 (2022)

Wu, K., Shen, J., Ning, Y., Wang, T., Wang, W.H.: Certified edge unlearning for graph neural networks. In: Proceedings of the 29th ACM SIGKDD Conference on Knowledge Discovery and Data Mining, pp. 2606–2617 (2023)

Wu, J., Yang, Y., Sui, Y. et al.: Gif: a general graph unlearning strategy via influence function (2023a). arXiv:2304.02835
XenonStack.: Guide to Enterprise Machine Learning and its Use Cases (2022). https://www.xenonstack.com/insights/enterprise-machine-learning-use-cases
Xiao, C., Ghosh, A., Wu, H., Wang, S., van Thiel, D.: Benchmarking federated machine unlearning methods for tabular data (2025). arXiv:2504.00921v1. https://arxiv.org/abs/2504.00921
Xu, J., Wu, Z., Wang, C., Jia, X.: Machine unlearning solutions and challenges. IEEE Trans. Emerg. Topics Comput. Intell. **7**(2), 223–245 (2023b). https://doi.org/10.1109/TETCI.2023.3273145
Xu, H., Zhu, T., Zhang, L., Zhou, W., Philip S.Y.: Machine unlearning: a survey. ACM Comput. Surv. **56**(1), 1–36 (2023a)
Xu, H., Jiang, S., Wang, J., Yu, H., Zhou, Q., Qian, H., Wang, B.: Machine unlearning: a survey. In: ACM Computing Surveys, vol. 55, no. 11, pp. 1–36 (2023).https://doi.org/10.1145/3603620
Xu, J., Xiao, Z., Zhang, Y., Wang, J., Li, X.: Machine unlearning: solutions and challenges (2023). arXiv:2308.07061. https://arxiv.org/pdf/2308.07061. Accessed 15 Sept 2025
Xu, X., Yue, X., Liu, Y., Ye, Q., Hu, H., Du, M.: Unlearning isn't deletion: investigating reversibility of machine unlearning in LLMs (2025). arXiv:2505.16831
Xu, H., Zhu, T. et al.: Machine unlearning: a comprehensive survey. ACM Computing Surveys (2023b)
Xue, L., Hu, S. et al.: A survey of machine unlearning verification, challenges, and future directions (2025). arXiv:2506.15115
Yan, H., Li, X., Guo, Z., Li, H., Li, F., Lin, X.: Arcane: an efficient architecture for exact machine unlearning. In: Raedt, L.D., (eds.) Proceedings of the Thirty-First International Joint Conference on Artificial Intelligence, IJCAI-2, pages 4006–4013. International Joint Conferences on Artificial Intelligence Organization, 7 (2022). https://doi.org/10.24963/ijcai.2022/556
Yan, H., Li, X., Guo, Z., Li, H., Lin, F., Chen, X.: An efficient architecture for exact machine unlearning. In: Proceedings of the Thirty-First International Joint Conference on Artificial Intelligence (2022). https://www.ijcai.org/proceedings/2022/0556.pdf. Accessed 14 Sept 2025
Yan, H., Li, X., Guo, Z., Li, H., Li, F., Lin, X.: Arcane: an efficient architecture for exact machine unlearning. In: De Raedt, L. (eds.) Proceedings of the Thirty-First International Joint Conference on Artificial Intelligence, IJCAI-22, pp. 4006–4013. International Joint Conferences on Artificial Intelligence Organization, 7 (2022). https://doi.org/10.24963/ijcai.2022/556. Main Track
Yan, H., Li, X., Guo, Z., Li, H., Lin, F., Chen, X.: An efficient architecture for exact machine unlearning. In: Proceedings of the Thirty-First International Joint Conference on Artificial Intelligence. https://www.ijcai.org/proceedings/2022/0556.pdf (2022d). Accessed 14 Sept 2025
Yang, R., He, M., He, Z., Qiu, Y., Huang, X.: MUSO: achieving exact machine unlearning in over-parameterized regimes (2024). arXiv:2410.08557. https://arxiv.org/abs/2410.08557
Yang, X., Wang, X., Zhang, Q., Petzold, L., Wang, W.Y., Zhao, X., Lin, D.: Shadow alignment: the ease of subverting safely-aligned language models (2023). arXiv:2310.02949
Yao, S., Neel, S., Zhang, R., Chen, Y., Yue, Y., Liang, P., Kolter, J. Z.: Mislabeled data vs. memorization: how does neural network respond to ambiguous training data? In: The Eleventh International Conference on Learning Representations (2023)
Yao, Y., Jia, J., Liu, S., Liu, B., Hoover, J., Cheng, Y.: Exploring the landscape of machine unlearning: a comprehensive survey (2023). arXiv:2305.06360
Yao, Y., Xu, Y., Liu, Y.: Large language model unlearning. In: The Thirty-eighth Annual Conference on Neural Information Processing Systems (2024)
Yin, R.K.: Case study research and applications: design and methods (6th ed.). SAGE Publications. https://ebooks.umu.ac.ug/librarian/books-file/Case%20Study%20Research%20and%20Applications.pdf (2018). Accessed 10 Sept 2025
Yu, S., Sun, F., Guo, J., Zhang, R., Cheng, X.: Legonet: a fast and exact unlearning architecture (2022). arXiv:2210.16023. https://doi.org/10.48550/arXiv.2210.16023
Yuan, X., Pang, T., Du, C., Chen, K., Zhang, W., Lin, M.: A closer look at machine unlearning for large language models. In: Proceedings of ICLR 2025 (2024). arXiv:2410.08109

Zhang, H., Nakamura, T., Isohara, T., Sakurai, K.: A review on machine unlearning. SN Comput. Sci. **4**(4), 337 (2023b)

Zhang, H., Nakamura, T., Isohara, T., Sakurai, K.: A review on machine unlearning. SN Comput. Sci. **4**(4), 337 (2023c). https://doi.org/10.1007/s42979-023-01826-6

Zhang, L., Zhu, T., Zhang, H., Xiong, P., Zhou, W.: Fedrecovery: differentially private machine unlearning for federated learning frameworks. IEEE Trans. Inf. Forensics Secur. **18**, 4732–4746 (2023d)

Zhang, H., Nakamura, T., Isohara, T., Sakurai, K.: A review on machine unlearning. SN Comput. Sci. **4**(4), 337 (2023e)

Zhang, D., Pan, S., Hoang, T., et al.: To be forgotten or to be fair: unveiling fairness implications of machine unlearning methods. AI Ethics **4**, 83–93 (2024a)

Zhang, D., Pan, S., Hoang, T., Xing, Z., Staples, M., Xiwei, X., Yao, L., Qinghua, L., Zhu, L.: To be forgotten or to be fair: unveiling fairness implications of machine unlearning methods. AI Ethics **4**, 83–93 (2024b)

Zhang, Y., Liu, Z., Zhou, L., Liu, Y., Chen,H.: Grapheraser: a general framework for machine unlearning in graph neural networks. In: ACM Conference on Computer and Communications Security (CCS) (2022). https://arxiv.org/abs/2208.11300

Zhang, D., Pan, S., Hoang, T., Xing, Z., Staples, M., Xu, X., Yao, L., Lu, Q., Zhu, L.: To be forgotten or to be fair: unveiling fairness implications of machine unlearning methods (2023)

Zhang, R., Lin, L., Bai, Y., Mei, S.: Negative preference optimization: from catastrophic collapse to effective unlearning (2024). arXiv:2404.05868

Zhang, Y., et al.: Graph unlearning. In: Proceedings of the 29th ACM SIGSAC Conference on Computer and Communications Security (CCS), pp. 1567–1580 (2022). https://doi.org/10.1145/3548606.3559352

Zhao, C., Qian, W. et al.: Rethinking adversarial robustness in the context of the right to be forgotten. In 41st International Conference on Machine Learning, volume 235 of PMLR, pp. 60927–60939 (2024)

Zhou, Z., Wu, Q., Tan, C., Wang, Z., Gu, S., Kwok, J.T., Yang, Q.: Federated learning: challenges, methods, and future directions. IEEE Signal Process. Mag. **37**(3), 50–60 (2020). https://doi.org/10.1109/MSP.2020.2975877

Zhou, C., Fu, A., Dai, Z.: TruVRF: towards triple-granularity verification on machine unlearning (2024). arXiv:2408.06063

Zhou, C., Li, N., Gao, Y., Chen, H., Fu, A., Zhang, Z., Yu, S.: Comprehensive survey on machine unlearning for large language models. In: Proceedings of the 2024 International Conference on Privacy-Preserving Machine Learning (2024)

Zhu, X., O'Connell, P.: Accountability in AI: Governance and Responsibility. MIT Press (2024)

Zou, A., Wang, Z., Carlini, N., Nasr, M., Kolter, J.Z., Fredrikson, M.: Universal and transferable adversarial attacks on aligned language models (2023). arXiv:2307.15043

The manufacturer's authorised representative in the EU is Springer
Nature Customer Service Centre GmbH, Europaplatz 3, 69115 Heidelberg,
Germany. If you have any concerns regarding our products, please
contact ProductSafety@springernature.com

Printed and bound by CPI Group (UK) Ltd, Croydon, CR0 4YY
17/09/2026
02218433-0010